Shuttle into Space

A Few Other Books by the Author

The Handbook of Model Rocketry
The New Model Rocketry Manual
The Third Industrial Revolution
The Lionel Train & Accessory Manual
Model Rocket Safety
The Model Rocketry Manual
Man and the Space Frontier
Earth Satellites: The Race for Space Superiority
Rocket Power and Space Flight

(As "Lee Correy")
Contraband Rocket
Rocket Man
Starship Through Space

Shuttle into Space

A Ride in America's Space Transportation System

G. Harry Stine

Full-Color Illustrations

Follett Publishing Company/Chicago

Acknowledgments

The author wishes to express his thanks to the following people from Rockwell International Space Division, Downey, California, for their helpful assistance and for providing the illustrations used in this book: Patti Mancini, B. J. Long, Paula H. Robb, William F. Green, and Al Ogram.

All characters appearing in this book are fictitious, and any resemblance to actual persons, living or dead, is purely coincidental.

ISBN 0-695-40920-4 Titan binding
ISBN 0-695-30920-X Paper binding

Library of Congress Catalog Card Number: 78-51262

Second Printing

To

MICHAEL COLLINS

pilot, astronaut, author

CONTENTS

Space Shuttle System

Top View

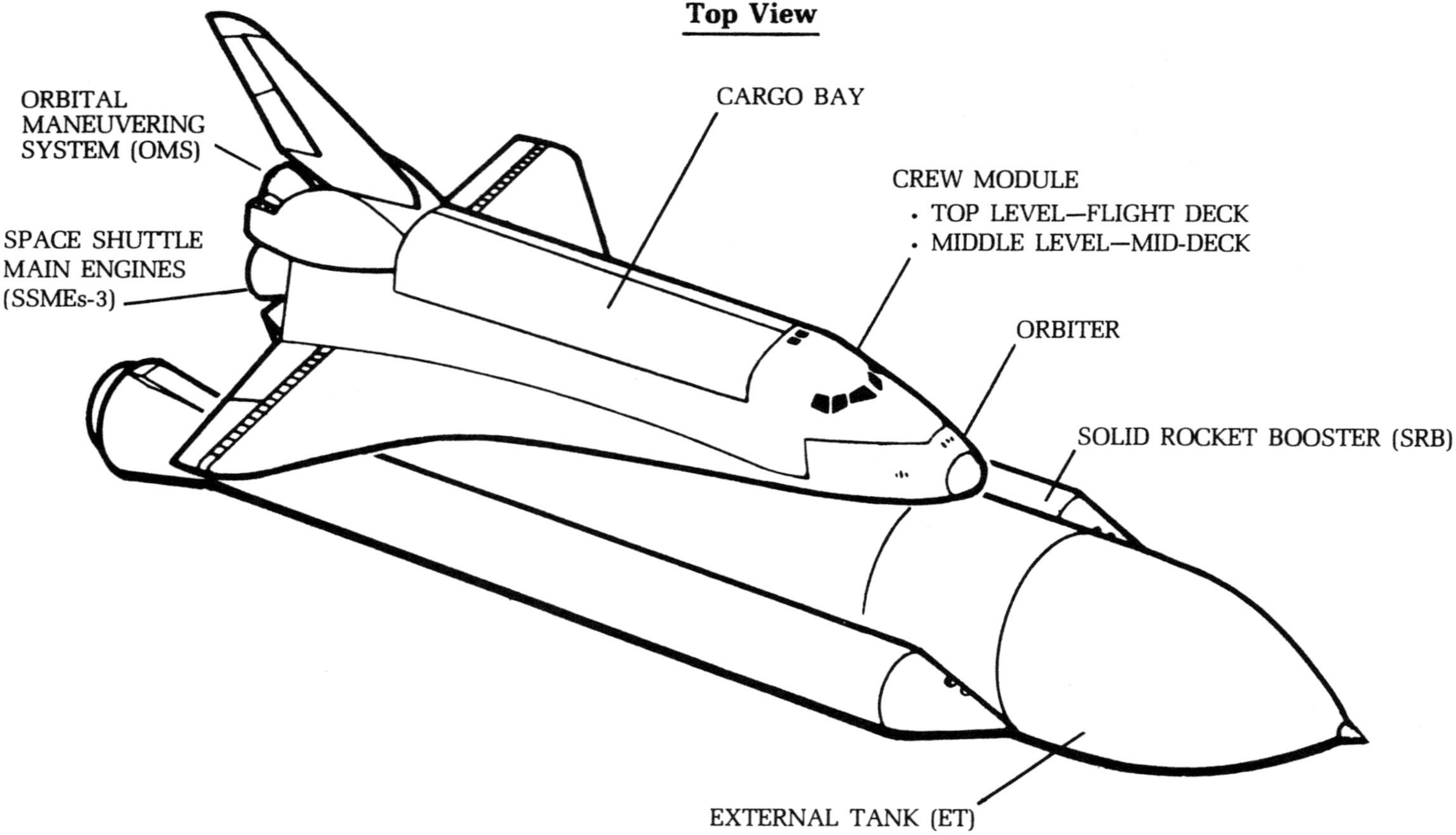

Space Shuttle System

Side View

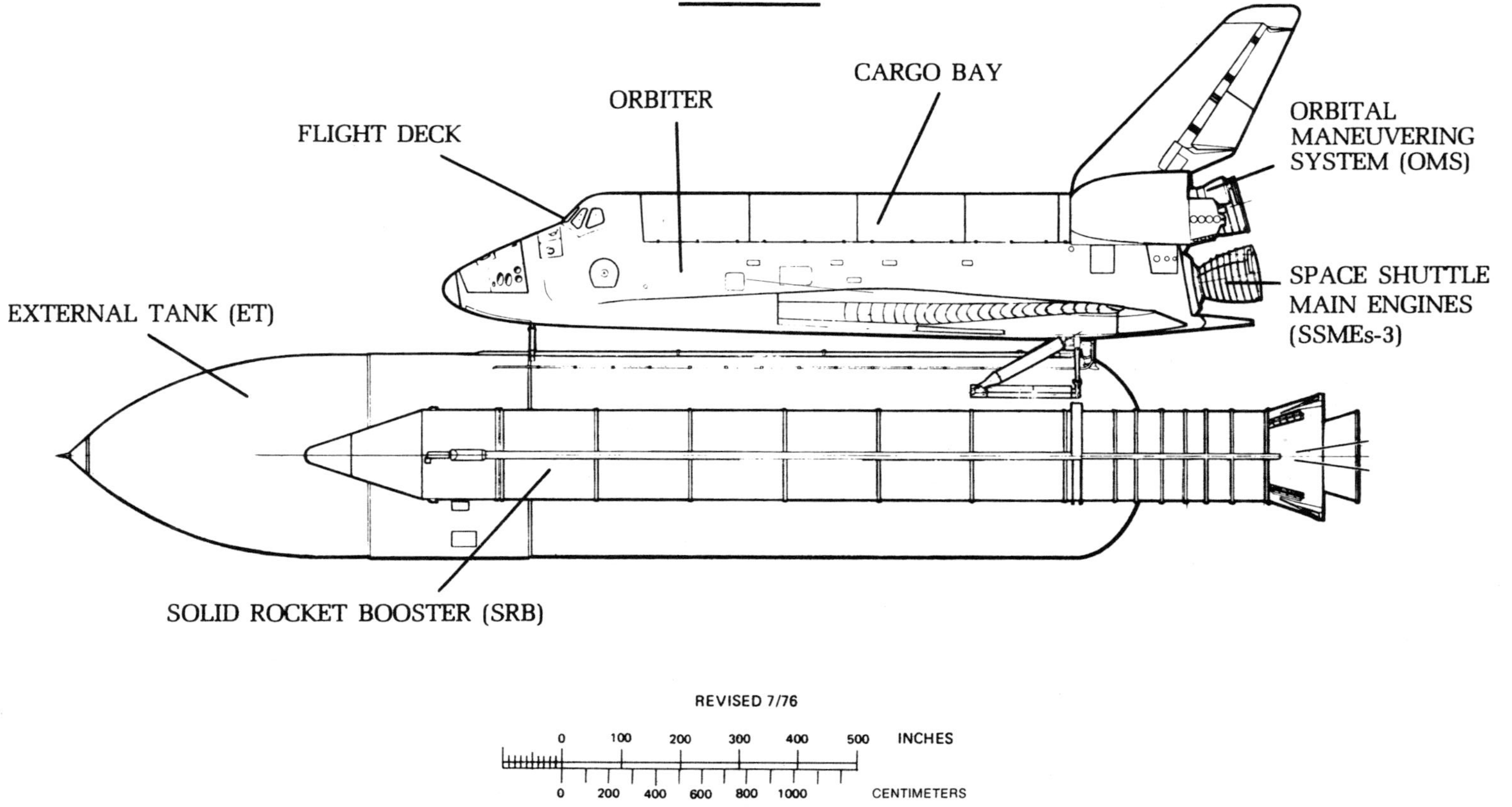

Space Shuttle System

Front View

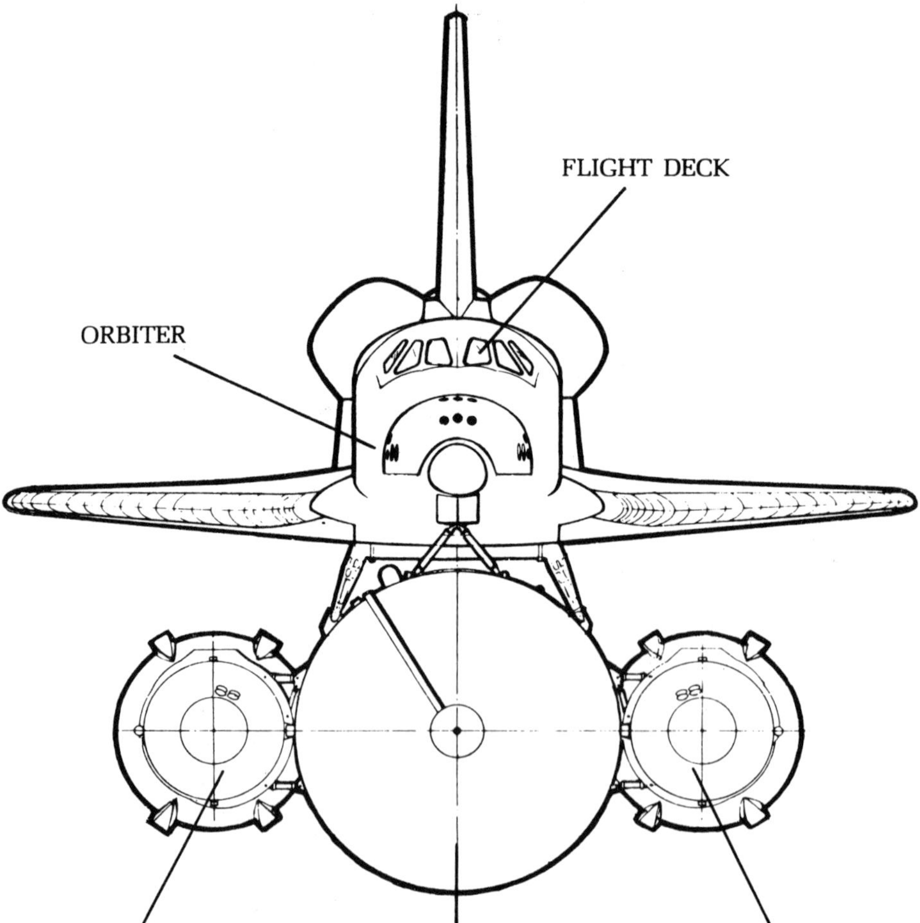

Space Shuttle System

Back View

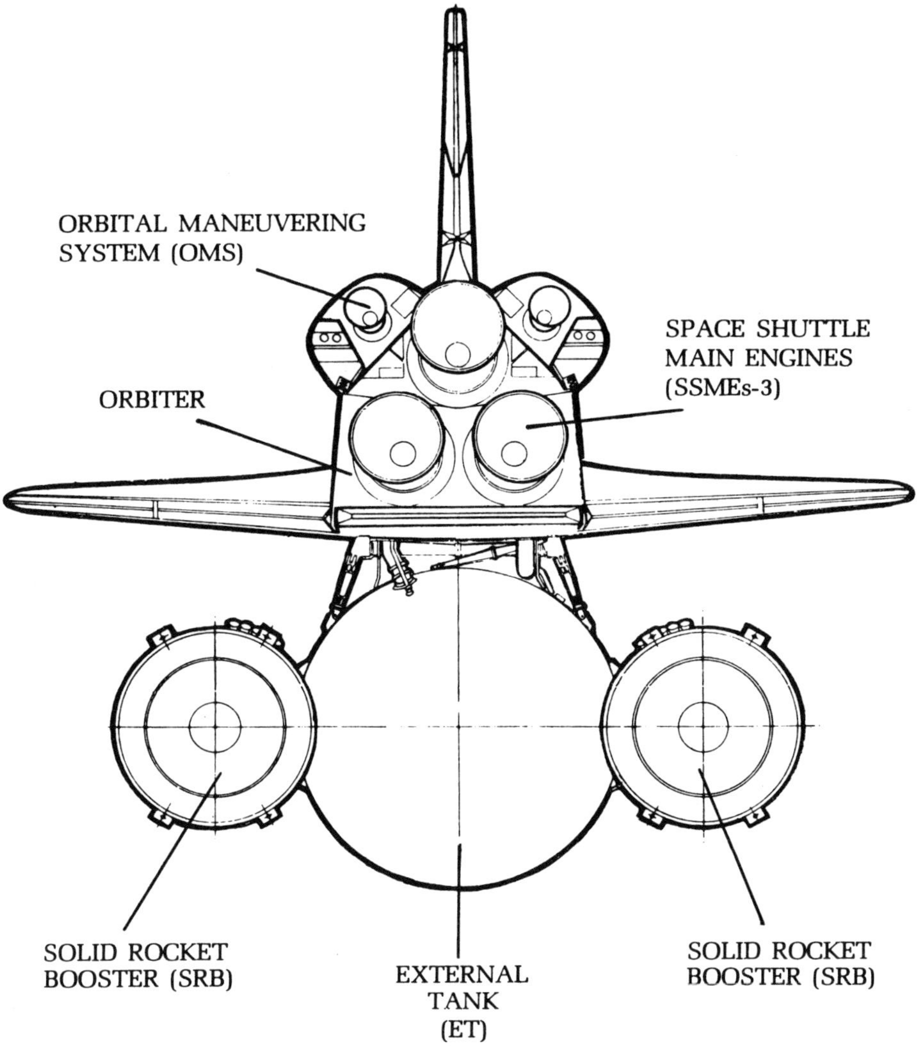

1 The Space Truck

It's waiting there for us on the launch pad at Cape Canaveral, Florida.

Poised over the gaping flame trench, the Space Shuttle Orbiter *Enterprise* is surrounded by its servicing towers, almost ready for this week's flight into earth orbit. Thirty-two tons of payload have been tucked into its fifteen-by-sixty-foot cargo bay. And tomorrow, when the *Enterprise* lifts off at 8:00 A.M., *we* will be tucked aboard, too.

In one corner of the orbiter's cavernous cargo bay is our space experiment. You and I have worked two years designing and building it. If our ideas are correct, we will be able to make space wood, a new high-strength material that can be produced only in the weightlessness of outer space.

The two of us are not astronauts; we don't know

how to operate the Shuttle or any other space vehicle. Astronauts who flew the earlier manned spacecraft—the *Mercury*, *Gemini*, and *Apollo* capsules—were all military or civilian test pilots with more than 3,000 hours of jet flying time in their logbooks. On top of this experience, they underwent long, arduous specialized training for space flight. And they were all in outstanding physical condition. You and I (and anybody else who rides in a Space Shuttle) are just ordinary people who have passed a simple physical examination and have undergone several weeks of basic training to learn how to live in space.

But the fact that ordinary people can fly in a Space Shuttle is only one of the things that make it a unique spaceship. Even though the *Enterprise* now sits on the same launch pad from which the *Saturn V* rockets sent astronauts to the moon, the *Enterprise* doesn't look a bit like the *Saturn V*. In fact, it is totally different from every space rocket that has ever been built.

Although it flies into space like a rocket, it comes back to earth and lands like an airplane.

It can be used over and over again.

It is a space truck designed to haul *thirty-two tons* of cargo into orbit and to bring back from space almost as much payload.

As a matter of fact, the *Enterprise* returned from orbit about two weeks ago. It landed on the long runway to the west of the Vertical Assembly Building (VAB) at the Kennedy Space Center. It has since been unloaded, checked out, and had a new payload installed in its cargo bay. It was mated, or connected, to its propellant sources, the External Tank (ET) and the Solid Rocket Boosters (SRBs), in the VAB, then rolled out to the launch pad as the old *Saturn* rockets were. Now it stands ready to fly into space again tomorrow.

The bark of a loudspeaker demands our attention as its echoing voice bounces around Launch Complex 39A. "All Cape personnel, stand by for the landing of the Space Shuttle Orbiter *Constellation*! The orbiter has completed entry and is turning to final landing approach twenty miles northwest of the Cape. Ground service equipment should move into assigned positions."

Just west of the huge VAB, we can see the landing runway, a strip of asphalt 15,000 feet long and 300 feet wide. It looks like other airport runways at regular jet terminals around the world. But this runway was built across the Florida marshlands for the Space Shuttle orbiters to land on.

At this point in its flight, the *Constellation* is an unpowered airplane, a glider, coming back to earth with its stubby delta-shaped wings supporting it. It is flying under the control of a fully automatic landing system just as the Douglas DC-10 or the Lockheed L-1011 *TriStar* jet airliners usually do. The pilot-commander and the copilot of the *Constellation* have manual controls similar to those in a jet airliner, and, if necessary, they can manually override the automatic system to bring the *Constellation* to a safe landing.

The *Constellation* isn't in sight yet, but the

running dialogue from the loudspeaker tells us what is going on. "*Constellation* is at twelve-thousand-feet altitude ten miles from the end of the runway. The landing system indicates that it has acquired and is functioning normally. On-board guidance and control system has switched to automatic landing mode. Glide slope is established at twenty-four degrees. *Constellation* is now aligned with the runway on glide slope at three hundred miles per hour. . . ."

Now we can see it! It's a tiny black dot high in the northwestern sky. It's coming fast and dropping quickly. Now it looks like a triangular-shaped black wedge. It gets bigger and bigger, until it's close enough to identify as the big, black-bottomed Space Shuttle Orbiter *Constellation.*

"Four thousand feet from end of runway . . . altitude two thousand feet . . . speed three hundred miles per hour. . . . Now starting initial flare. . . . Altitude five hundred feet . . . landing gear down. . . ."

Just like a jet airliner, the *Constellation* lowers its tricycle landing gear. It approaches the runway steeply with its nose at a high angle like the *Concorde* SST. Although it is moving at almost 200 miles an hour, it is so big that it appears to be floating over the runway.

"Over the runway threshold . . . two hundred miles per hour. . . . Final landing flare. . . . Touchdown!"

With a puff of smoke, the main landing gear tires kiss the runway surface. The big vertical rudder opens up as an air brake. Down comes the nose until the nose-gear tires touch the runway with another puff of smoke. The *Constellation* is now rolling along the runway just like any airliner landing at the local airport.

But the *Constellation* isn't an airliner from another airport. It is a spaceship that has returned from orbit.

It rolls to a smooth halt long before it reaches the end of the three-mile runway. A throng of cars and trucks converges on it. A truck with a vertical-lift bed backs up to the side of the *Constellation*'s nose. A circular hatch opens, and seven people get out. Four of them are regular crew members. Three are scientists.

The *Constellation* has just spent seven days in orbit around the earth. It has surveyed the mineral, farm crop, and timber resources of Brazil, New Zealand, and the northwestern United States. It has helped survey for new petroleum resources in Australia, and has made measurements of the ozone layer around the earth.

The excitement of the landing fades as routine ground work begins. The *Constellation* will be towed off the runway and over to the VAB. There, its cargo bay will be unloaded. It will be carefully inspected. Technicians may replace a few of the 34,000 special ceramic tiles that cover its surface and protect it against the blazing heat of entry into the earth's atmosphere from orbit. A new payload will be installed in the cargo bay, which is big enough to hold an entire railroad boxcar. Then the *Constellation* will be hoisted to the vertical position and mated with its External Tank and Solid Rocket Boosters. When everything is ready to go, in about two weeks, it will be wheeled out to the launch pad on the mobile launcher and sent into orbit. Its flight may last from one to thirty days, depending on its mission and the work to be performed on board.

"All personnel in Launch Complex 39A, clear the launch area. Loading of hypergolic propellants will begin when the pad is clear of unauthorized personnel," warns the loudspeaker. The crews servicing the *Enterprise* are ready to load the nitrogen tetroxide oxidizer into the tanks of the rocket engines in the Orbital Maneuvering System (OMS), the rocket system used to move the orbiter in space. Nitrogen tetroxide is a toxic, or poisonous, chemical, and special equipment must be worn by the crews when fueling the tanks. They will be finished with the fueling in about an hour, and it will then be safe for us to board the *Enterprise* for a final check of our equipment in the payload bay.

While we wait, we decide to explore the VAB. The Vertical Assembly Building at Kennedy Space Center is still the world's largest building. It is 525 feet high and covers eight acres. Originally designed and built in the 1960s to assemble the *Saturn* moon rockets, it has been remodeled for use by the Space Shuttles. Although the Shuttles, when ready for launch, are only 184 feet tall, their 78-foot wingspan required the big VAB doors to be widened. Adaptations were also made on the mobile launchers and transporters used for the *Saturn* rockets so they could be used to launch the Shuttles.

All of this remodeling and reuse of facilities has kept down the cost of the Space Shuttle program. It has cost less than one-fifth of the *Apollo* lunar landing program of 1968 to 1972.

One thing about the VAB has not changed since its construction. It is difficult to comprehend the size of the building. As we drive the three miles from the launch pad to the VAB, the building doesn't seem to get any bigger. It is so large to begin with that we have nothing in our experience with which to compare it, and since it sits all by itself in the middle of flat marshlands, there are no other large structures around it for comparison.

Inside the VAB, the Space Shuttle *Constitution* is being prepared for its flight next week. Cranes are hoisting the *Constitution* orbiter into position on its stack, the already assembled ET and SRBs. Inside the closed cargo bay is the ESA Spacelab.

Spacelab is an international scientific project. It was designed and built by the European Space Agency, a consortium made up of Belgium, Denmark, France, Ireland, Italy, the Netherlands, Spain, Switzerland, the United Kingdom, and West Germany. Spacelab is a scientific laboratory for studying the earth, the solar system, the stars and galaxies, and other aspects of space. It is a module designed so that its various elements can be changed to suit the requirements of scientists and engineers who wish to carry out experiments in space. The desired elements are simply plugged into it as you would plug a new component into your stereo music system, and, in turn, the entire Spacelab is plugged into the cargo bay of a Space Shuttle orbiter.

Because the Space Shuttle system allows ordinary people to ride it, scientists and engineers have been able to fly into space to conduct their Spacelab experiments in person. The scientific results from Spacelab in the few years it has been flying have already filled several books—and caused many others to be rewritten. Spacelab is enabling us to learn more about our home planet and our universe—learning that is very important because earth is the only planet we have to date and because we are children of the universe.

The knowledge provided by Spacelab and other experiments has been made possible by the fact that Space Shuttles are like airplanes: they can be reused after each flight. Earlier rockets were one-shot affairs. They were launched once and then discarded into the ocean or outer space, leaving only their payloads to continue to work in space. Naturally, this was a very wasteful and expensive procedure, rather like building an ocean liner, sailing it once across the ocean, and sinking it upon arrival.

The Space Shuttle is the first space transportation system. It is like a car, a train, a boat, or an airplane, for it can be used over and over again. It makes flying into space easier and cheaper. You and

I could not make a trip into orbit aboard a *Saturn V* because it cost $125 million to launch each one of them. To fly a Space Shuttle mission into orbit and back costs about $20 million—and the cost is coming down all the time because the National Aeronautics and Space Administration (NASA) is learning more efficient and less expensive ways to repair, launch, operate, and land the Space Shuttles.

"*Enterprise* crew, report to Launch Complex 39A for final preflight inspection," the loudspeaker crackles, echoing throughout the huge VAB.

That's us. We are scheduled to make the final inspection of our equipment aboard the *Enterprise*. Quickly we leave the VAB, get in our car, and ride out to the *Enterprise* launch pad.

2 On the Pad

The *Enterprise* and its ET and SRBs are completely shrouded by the service towers and payload change-out room, previously called a white room, which is super-clean to prevent any dirt from entering the *Enterprise* systems and causing trouble later. The towers permit people to work on various parts of the *Enterprise* stack as it sits vertically on the launch pad, but they will all roll back on railroad tracks before tomorrow morning's liftoff.

The tower elevator lifts us toward the *Enterprise*'s nose and deposits us at the level of the cabin entrance hatch. We're the last to arrive. The other five people on the flight are already on hand.

The chief launch controller is standing by for the final crew briefing. He is Sam Perona, a fifteen-year veteran of NASA manned space flight operations at Kennedy Space Center. He helped launch the *Saturn-IB* rockets for the Skylab and *Apollo-Soyuz* missions of the 1970s, and he has been in charge of launching more Space Shuttle missions than any other person at the cape. We're fortunate to have him as our chief launch controller because he is in direct charge of the ground crew assigned to the *Enterprise*, and both he and the crew have followed the ship through its most recent refurbishment.

Standing quietly next to Perona is our pilot and commander, Richard Edwards. He is experienced at his job, too. For many years, he flew experimental jet aircraft for NASA's Dryden Flight Research Center at Edwards Air Force Base in California. He has flown over a dozen Shuttle flights. Our copilot, David McQuire, is also an experienced jet and Shuttle pilot. Both men are in their mid-thirties, and of all the Space Shuttle people, they are the most highly trained and the most like the old astronauts. The two of them have almost the same qualifications as *Apollo* astronauts—at least 3,000 hours of jet flying time, university degrees in engineering, and outstanding physical conditions.

Our third crew member, payload specialist Khambrel Johnson, is not a jet pilot, but he is a licensed pilot of non-jets. He, too, is highly trained, holds a degree in engineering, and is in good physical condition. Although Richard Edwards is commander of the *Enterprise*, Johnson "flies" the ship once it is in orbit. He is responsible for all on-orbit operations, such as maneuvering the orbiter for rendezvous with other objects in space.

Working with him will be our mission specialist, Mary Howard. Her job is supervising the safety of the *Enterprise* in orbit because she maintains the services the orbiter supplies to its payload. She oversees the electrical systems, cooling systems, and pressurized gas systems.

Our companion on the mid-deck is Dr. Barbara Cleveland, a scientist with the National Oceanic and Atmospheric Administration (NOAA), who will be conducting an experiment using a laser to measure the height of waves on the earth's oceans. She'll also use her equipment to determine the distance between the crests of waves and to plot their direction. Later, she'll compare the information she obtained aboard the *Enterprise* with weather satellite photographs and surface weather observations, all made at the same time her laser wave measurements were made in orbit. Her work is part of a long-range research project to determine the effects of oceans on both weather and climate.

Sam Perona begins our final prelaunch briefing. It's basically a reminder session because we've all had some previous training at the Johnson Space Center outside Houston, Texas.

"Ladies and gentlemen, even though you've been through all this before. . . . How many flights for you now, Mary? Four, is it? And five for Khambrel, if I recall. And more than a dozen for you two old hands, right?" He grins at Edwards and McQuire, who nod and grin back. "Well, as I said, even though you've been through it before, we can't afford to get stale or lazy. So let's review again the details from now until launch.

"When you go aboard in a few minutes for your final checks, I know you'll be most interested in the equipment you'll be responsible for during the mission. Make sure that all of your equipment listed on the manifest is aboard and in its assigned place, if you can get to it. After that, make sure any personal items are aboard and in place. Yes, we have ground crews to check this, but *you* are the ones who'll be flying in this brick spaceship tomorrow. Check it out yourself to make sure. I won't bring the *Enterprise* back after launch if you've forgotten something. Once you leave the cabin tonight, nobody will enter it except the two check-out pilots."

The "manifest" that Perona referred to is a detailed, itemized, computer-generated list of all items aboard the *Enterprise*, down to the last moist towelette and the Band-Aids in the first-aid kits. NASA isn't as strict as they used to be about material brought on board; now they will let crew members like ourselves take along personal belongings such as books, mementos of the flight, and other small items. Since the Shuttles started flying, space travel has become more commonplace, and many more people have traveled into orbit. It's no longer the exclusive experience of a few select people. Early worries about some crew members selling first-day covers, medallions, and other souvenirs that had been taken into space in the Space Shuttles have evaporated. With regular weekly launchings taking place, orbital flight has become almost as exclusive as a rainstorm, so flight souvenirs have limited commercial value. However, like most people, we want to commemorate our own first flight into space, and so in our personal equipment lockers on board we've stowed a few mementos that we want to put on our bookshelves when we get home again.

But Sam Perona is continuing his briefing. "When you leave the launch pad tonight, you'll go directly to the transient crew quarters, where you'll be isolated for the last twelve hours before liftoff to make sure you haven't caught a cold or something worse. Getting sick in orbit is no fun, and you could jeopardize the whole mission if you became sick enough to require an emergency landing.

"After dinner in your quarters this evening, your time is your own, but I must ask you not to leave your quarters or to entertain guests. There are telephones if you want to call anyone—and a TV if you want to look. There's also a computer terminal we can hook up by telephone with any other computer in the country if you want to run some last-minute calculations. But I suggest that you be in bed before ten tonight. The steward will give you a sleeping pill if you're too excited to sleep. We want you completely rested for your flight so you can perform your duties and conduct your work at your best. Although flying into orbit is fairly commonplace today, we don't want you to waste a minute of your flight if we can help it."

The briefing is just that: brief. We follow

Edwards, McQuire, Johnson, and Howard through the forty-inch-diameter crew entrance hatch on the left side of the nose of the *Enterprise.*

Things look different. We've spent hours in the Space Shuttle training simulator in Houston, going through simulated flights while remaining on the ground. And we've been in the *Enterprise* back in the VAB, installing our equipment. But the simulators and the check-outs in the VAB were somehow different. Although the *Enterprise* is laid out like the simulators, it is not *exactly* like them. The bricklike exterior shows the effects of many entries into the earth's atmosphere; the ceramic tiles have a hot, burned look to them, and there are flow lines of brown and black soot streaming aft over the nose of the orbiter where the ceramic thermal protection system has boiled away in the blazing heat of entry. The inside also looks different from the simulators. The walls and floor have a scuffed appearance. People have lived in this cabin.

In addition, it is always somewhat of a surprise to climb into the *Enterprise* in its vertical position. We seem to think of the Space Shuttle orbiter as it rests horizontally on its landing gear. But here it is in the takeoff position. What was the rear wall of the mid-deck is now the floor. Our seats are now on the wall, and the forward equipment lockers are on the ceiling.

Removable aluminum ladders enable us to get to our seats and to climb up to check the personal equipment lockers on the forward bulkhead, now temporarily on the ceiling. We cannot open some of the lockers now because the contents would spill out. We just have to trust that the crew did not remove the change of underwear we each put in there several days ago.

Some of our experimental equipment is housed in the aft, or rear, lockers, which we are able to check because they're now on the floor. Afterwards, we clamber down the ladder leading through the open air lock into the cargo bay. Temporary lights have been rigged to illuminate the bay; they will be removed later tonight. The payload bay doors are closed and cannot be opened on the ground without special supports. To save weight, the doors were designed and built to operate and support themselves only in the weightlessness of space. The special hoists and rigging required to open the doors on the ground are now stowed away in the service towers.

The remotely operated equipment for our experiment is located in the left front corner of the cargo bay (as viewed with the *Enterprise* in a horizontal position). Special ladders and walkways have been temporarily installed so we can get to our equipment. A quick check and a short test indicate that the equipment is ready to fly.

We go back up the ladder through the air lock to the mid-deck, then on to the flight deck.

The *Enterprise* has a nose section with three decks, one above the other when the ship is horizontal. The lowest deck contains the environmental systems that maintain cabin air pressure, remove the carbon dioxide exhaled by crew members, eliminate odors, and keep the cabin air at a comfortable temperature and humidity.

The mid-deck—where we will ride with Dr. Cleveland—is the general living quarters. It contains three seats, sleep arrangements, a personal hygiene center, equipment lockers, the air lock leading to the cargo bay, the food galley, a foldaway table, and the zero-gravity toilet, a device which NASA refers to as the "waste management facility."

The top deck is the flight deck and looks very much like the cockpit of a Boeing 747. The two pilots ride side by side looking out through an airliner-type windshield. Behind the two pilots ride the payload specialist and the mission specialist. The side walls and the rear bulkhead of the flight deck contain the equipment controls for operating the *Enterprise* in orbit. The controls for our experiment are located on the right side of the aft bulkhead.

It's very difficult to get to the control panel for our experiment because the flight deck's rear bulkhead is now down, forming a temporary floor. Mary Howard is kneeling on the blank panels below the operating consoles, testing the various environmental systems. There is just enough room for us to join her.

Our control console lights up when we put electrical power to it. A quick ten-minute test tells us that our experiment is properly installed and ready to go. We crawl back to the mid-deck and out through the forty-inch hatch to the white room in the service tower.

Edwards and McQuire are there, going over the final additional items on the manifest with Khambrel Johnson. We tell them we're all set to fly and are returning to the transient crew quarters for the night.

Our NASA car is waiting at the foot of the service tower to take us to our quarters at Kennedy Space Center several miles away. Our quarantine begins as we enter. The quarters are comfortable and attractively furnished like a first-class motel. Every person who comes in contact with us from now until launch time is "clean." Each has passed a physical check and has no contagious diseases, not even a case of the sniffles. Nor has any of them been with anyone who has had even the sniffles for the past week. These extra precautions pay off. No Space Shuttle has ever had to make an emergency landing because of crew illness.

The rest of our crew show up one by one. We have a good dinner together. There is little talk about the flight tomorrow. Mostly, we get to know one another better. We all know we have to do a job in orbit. The crew that flies and operates the *Enterprise* is well-trained and competent, and they are easy people to get to know and work with. You and I are very impressed and glad we'll be flying with them.

After dinner, we turn on the television. A rerun of a "Star Trek" episode comes on. Usually I can't resist the adventures of Captain Kirk and his crew. But tonight the show seems pale. After all, tomorrow morning we're going to be doing the real thing on the real *Enterprise*, primitive as it may seem in comparison to traveling about the galaxy in a fictional starship.

It's a long and boring evening. Everything is ready. There's nothing to do now but wait—and try to get a good night's sleep.

This aerial view of the Space Shuttle launch area at Kennedy Space Center, Cape Canaveral, Florida, shows Launch Complex 39 with the Vertical Assembly Building (VAB) in the distance and the two launch pads used in the *Apollo* program in the foreground. A new runway to land the Shuttles has been built beyond the VAB. A Space Shuttle is riding out to Launch Complex 39B on a mobile transporter.

This exterior view of a NASA Space Shuttle shows the delta-winged orbiter atop the big External Tank (ET), with the two Solid Rocket Boosters (SRBs) on each side of the ET. The orbiter is covered with 34,000 ceramic tiles to provide thermal protection during the blazing entry into earth's atmosphere from orbit.

This cutaway drawing shows some of the internal arrangement of a Space Shuttle. The three Space Shuttle Main Engines (SSMEs) in the tail of the orbiter burn liquid hydrogen and liquid oxygen stored in the big External Tank beneath the orbiter. The three- or four-person crew rides on the upper flight deck of the orbiter like an airliner crew. The huge cargo bay is fifteen feet in diameter and sixty feet long, big enough to hold a railroad boxcar.

The Space Shuttle Orbiter *Constitution* is hoisted to mate with its External Tank and Solid Rocket Boosters, already poised on a mobile launcher inside the Vertical Assembly Building. The *Constitution* returned from orbit less than two weeks ago. The SRBs were recovered in the ocean from a Shuttle flight several months ago and have been reloaded with solid propellant. Only the big ET is expended on a flight.

Liftoff! With the two Solid Rocket Boosters and the three Space Shuttle Main Engines in the tail of the orbiter producing over 6,000,000 pounds of thrust, the *Enterprise* lifts off Launch Complex 39A on the start of its voyage into earth orbit and back.

Rocket engines roaring, the Space Shuttle *Enterprise* now clears the launch pad towers and climbs skyward, carrying thirty-two tons of payload and seven people into earth orbit.

Five thousand feet up, the *Enterprise* begins to tilt eastward under the commands of its automatic guidance system. Flight acceleration is less than 3 g's, or three times the force of gravity, making it possible for many people to ride the Shuttle without stress.

On the flight deck of the *Enterprise*, the regular flight crew rides into space. The two pilots, Richard Edwards and David McQuire, are seated in front with airliner-type controls within reach. Mission specialist Mary Howard and payload specialist Khambrel Johnson ride behind them.

3 Liftoff

The steward is gently asking me to wake up and get dressed. The room lights come on. The digital clock says 5:00 A.M.

I tumble out of bed and take a quick shower, the last I will have until we get back to earth. The *Enterprise* does not have a zero-gravity shower stall as the old Skylab did. We'll be able to freshen up with moist towelettes during the flight, but no more showers until we land.

It doesn't take long to dress. Unlike earlier space vehicles, the Shuttles provide a shirt-sleeve environment for their crews. Bulky space suits are unnecessary. Crew members dress in a normal type of uniform—white NASA flight coveralls made of fireproof beta cloth. Although the cabin atmosphere of the Space Shuttles is a standard atmosphere of 20 percent oxygen and 80 percent nitrogen at normal sea-level pressure, fire is still a worry in space flight. Nobody has been killed by fire in space since the unfortunate *Apollo-1* fire in 1967, and NASA is seeing to it that nothing like that ever happens again.

I stuff my eyeglasses and a ball-point pen in my coverall pocket. Other than small items like that, neither of us will take into space any of the things we normally carry in our pockets, such as keys, wallets, money, or pocketknives. There's no reason to have any of these things in space.

Breakfast is a full meal for everyone. We don't have to eat a low-residue diet as earlier space travelers did because there are toilet facilities aboard the *Enterprise*. The NASA crew is alert and eager in a subdued, professional way. I don't know about you, but I am so excited that it's difficult to concentrate on eating breakfast.

At 6:00 A.M., we're at Launch Complex 39A, boarding the elevator for the ride up to the white room in the service tower. The *Enterprise* seems alive now; it creaks, groans, and hisses. It's ready to fly except for pumping the liquid hydrogen and liquid oxygen into the big External Tank attached to its belly.

In the white room, we don our Snoopy helmets, the black-and-white cloth helmets similar to those worn by the *Apollo* astronauts. The helmets with communications earphones and microphones built into them have proven to be the best way to wear this equipment in space.

Finally, we are ready to go aboard. In NASA terminology, the mission schedule calls for "crew ingress."

Pilot-commander Dick Edwards and copilot Dave McQuire step through the hatch first because they must be strapped into the forward seats on the flight deck. They are followed by Mary Howard and Khambrel Johnson, who will ride in the flight deck seats behind the pilots. Then Dr. Cleveland goes aboard because she will occupy the right-hand seat on the mid-deck. Since I'm assigned to the middle seat, I go through the hatch next, and you follow. You have the left-hand seat opposite the only window on the mid-deck, a ten-inch-diameter porthole in the entrance hatch. Lucky you!

A ladder, which will be removed before launch, permits me to climb into my seat in the center of the mid-deck just forward of the air lock compartment. Climbing into a seat that's sticking out from what is now a wall is not an easy thing to do, especially when one must lie down in it. I get some help from the crew ingress technician of the ground crew. When I'm finally in place, I remove the plastic slippers that covered my velcro-soled flight shoes.

You are right behind me at this point, being helped into the left-hand seat and going through the same preflight strap-in procedures.

The restraint harness I don is almost identical to that used by the United States Air Force for almost thirty years. It consists of a lap belt, two shoulder belts, and a crotch strap. They are all connected to a common quick-release clasp over my chest. The lap belt holds me in the seat. The shoulder belts prevent me from doubling up over the lap belt. And the crotch strap keeps me from sliding out under the lap belt and shoulder belts. The ingress technician plugs in my communications headset so I can talk to the crew and hear the countdown. And then I'm all ready to go.

Early flights of the Space Shuttles weren't so simple. NASA wanted medical telemetry for every person aboard, which meant people had to ride with medical electrodes attached to their arms, legs, and chest so that NASA experts on the ground could monitor their heartbeat, breathing, pulse, and blood pressure. It was very quickly learned that this wasn't necessary. "Wall-to-wall data," one medical man called it. People can undergo a flight in a Space Shuttle with no ill effects, and after the first few flights, all the medical data began to look alike. So the requirement was dropped.

The ingress technician pats me on the shoulder, removes the access ladder, and ducks out the hatch with it. I turn to see you in your seat slightly behind me.

The ground crew swings the cabin hatch shut. We see the handles turning as it is sealed from the outside.

Now we have another long wait. Fortunately, our seats are comfortable. Amazing as it may seem, some people have been known to take a quick nap during this prelaunch period. I can't imagine how they were able to do it. My heart is racing with excitement.

The servicing towers must be rolled clear of the *Enterprise* before launch. Daylight streams through the tiny hatch porthole as the white room is wheeled back from the side of the ship. Soon we can see the northwest horizon and the flatlands of Cape Canaveral. But it's a somewhat restricted view.

Final loading of the liquid hydrogen and liquid oxygen into the External Tank begins. Since these propellants are very cold and boil away, they are the last things put aboard the Shuttle.

Liquid hydrogen is first loaded into the cavernous rear tank of the ET. A total of 378,378 gallons is pumped aboard. Liquid hydrogen is very cold stuff—minus 423 degrees Fahrenheit.

Even though the crew cabin of the *Enterprise* is a separate structure isolated from the main structure of the ship, like an egg riding in an egg carton, we can still hear and feel what is happening in the *Enterprise.* The extreme cold of the liquid hydrogen causes the materials of the ET and the *Enterprise*'s main engine bay to contract, or shrink. Engineers made careful allowance for this shrinkage when they designed the *Enterprise*, and many tests were conducted to check the design calculations. So there is really nothing to worry about even though the cooldown sounds terrible. Some of the creaks and groans of the structure can be felt as well as heard, and there is a muted hissing from the gaseous hydrogen that's boiling off and venting from the ET. The ET walls are hot to the super-cold liquid hydrogen, so the liquid hydrogen literally boils as it fills the tank and begins to cool down the structure. Eventually, the tank walls become cool enough so that the liquid hydrogen boil-off is held to a minimum.

Then 139,623 gallons of liquid oxygen (LOX) are pumped aboard. The ET is cooled down by now, so there isn't as much creaking and groaning. But we can hear and feel large volumes of liquids rushing through the loading valves and into the pipes at the rear of the *Enterprise.*

The liquid hydrogen and the LOX are fed through two seventeen-inch-diameter pipes to the three Space Shuttle Main Engines (SSMEs) located in the tail of the *Enterprise*. The pipes run between the *Enterprise* and the ET with quick-disconnect fittings so that the ET can be discarded in orbit. Most of the valves and control equipment is housed in the *Enterprise* itself. A minimum amount of equipment is located on the ET because that is an expendable part of the system. The ET is just a great big aluminum tank assembly 154 feet long and 28 feet in diameter. It is the only part of the Space Shuttle system that is used just once, and it is the space-going version of the drop-tanks used on long-range military aircraft.

The size, complexity, and cost of the *Enterprise* were drastically reduced by designing the system with an external and expendable liquid propellant tank. In addition, the ET has some uses in space, as we'll see later. Still, a few of the newer Shuttle designs now on the drawing boards have eliminated the need to discard anything; they are completely reusable single-stage-to-orbit vehicles.

The earphones in our Snoopy helmets bring us the chatter of the countdown communications network. Launch Control in the VAB directs the launch until the *Enterprise* clears the launch pad. After that, Mission Control in Houston takes over. We hear Edwards and McQuire talking with Perona in Launch Control.

"T-minus three minutes and counting. Stop topping liquid oxygen tank. Begin final topping of liquid hydrogen. Final guidance system alignment. Pilot enter Instruction 2464 into DISKY guidance computer."

"Launch Control, this is *Enterprise.* Instruction 2464 into DISKY. Computer accepts. Platform alignment complete. Standing by to copy final pad."

"Final pad coming at you." Launch Control then reads off a string of numbers that amount to the final instructions for the guidance and control computers that will steer us into orbit. Then, "T-minus two minutes and counting. Start pressurization of LOX tank. Close ET LOX valve."

"Launch Control, *Enterprise.* LOX vent valve indicates closed."

"Roger. Stand by for LOX pressurization complete. Telemetry to internal power. Guidance and control to internal power. Hydraulic pressure normal. Nitrogen purge complete."

"Launch Control, *Enterprise.* Roger. We copy

and confirm payload nitrogen purge complete. Nonpressurized compartment vents open. Fuel cell electrical bus voltages nominal."

"*Enterprise,* Launch Control. T-minus one minute and counting. LOX pressurization complete. Liquid hydrogen pressurization complete. All systems go for launch."

"*Enterprise.* Roger your last. All prelift checklists complete. All systems go for launch here."

From somewhere aft, we can hear and feel the banging of valves and the singing of pressurized systems. We are in the final long minute of the countdown. It is always the longest minute of any space operation. Back at Peenemünde, Germany, during World War II, when Wernher von Braun was developing the V-2 rocket, engineers called it the Peenemünde minute. It hasn't changed since then.

My earphones bring me the voice of our pilot on the intercom system. "Mid-deck, this is Edwards. You people all ready down there, I presume?"

Dr. Cleveland nods, licking her lips. You nod to me, too. I reply, "Roger for all three of us!"

"Okay, hang on. The first step's a dilly!"

"T-minus twenty . . . nineteen . . . eighteen . . ."

The tension of a space flight countdown has never changed.

"Fifteen . . . fourteen. . . . Ignition sequence start! . . . eleven . . . ten . . . nine. . . . Ignition of SSME! . . ."

The *Enterprise* shudders. From aft, we hear a muted thunder and feel a vibration. The three SSME rockets have ignited, drawing their liquid hydrogen and liquid oxygen from the big ET. It takes several seconds for them to build up to a full thrust of 1,125,000 pounds. But that's not enough to lift the 4,400,000 pounds the *Enterprise* weighs.

"Six . . . five . . . four. . . . SRB ignition! . . ."

We feel another shudder. The two huge solid propellant rocket boosters, each 12 feet in diameter and 149 feet long, ignite and come up to a thrust of 2,900,000 pounds each. Now the total *Enterprise* rocket thrust is 6,925,000 pounds.

The *Enterprise* roars off the launch pad in a holocaust of flame and smoke.

"Liftoff! We have liftoff!"

We feel a steady, increasing pressure that forces us into our seats. With about 1.5 g's acceleration, or a pressure one and a half times that of earth-normal gravity, the *Enterprise* rises vertically into the Florida sky.

"Clear of the tower! Roll program in. Pitch program in. Control transfer to Houston. Good luck, gang!" Perona signs off, transferring flight control over to Mission Control at the Johnson Space Center outside Houston.

As the *Enterprise* clears the launch pad towers, the guidance and control computer instructs the automatic pilot to roll the *Enterprise* to its flight direction. The autopilot causes hydraulic pistons to move the SSME and SRB rocket nozzles, changing the thrust direction, which in turn changes the flight direction. The *Enterprise* rolls so that its wings are in a roughly north-south line. And its nose pitches slightly up, or eastward, so that the ship begins to fly belly up.

Because there is no "up" or "down" in space, the up and down of the *Enterprise* is used as a reference by the computers and the crew. The top side of the orbiter is the side with the cargo bay doors.

There is a slight decrease in the acceleration pushing us into our seats. The SRBs have automatically decreased their thrust as they have been designed to do. This keeps the *Enterprise* acceleration below 3 g's, or three times the force of gravity, to prevent the ship's structure from being overstressed and to make the ride more comfortable for the human beings aboard.

"T-plus one minute. Thrust reduction achieved. Going right down the pike!" Mission Control reports.

"Roger from *Enterprise.* Looks copacetic from here, too. Nice ride today."

You can see out of the little ten-inch-diameter porthole in the hatch better than I. The surface of the earth is falling away. The sky is still blue, but it is a very dark blue as one would see it from a high-flying jet airliner. We're above the clouds, but haze makes it difficult to see any details on the surface below.

The *Enterprise* goes into orbit on its back. But it doesn't make any difference to us. The 3-g acceleration keeps us pinned to our seats, making us weigh three times our normal earth weight. Yet this acceleration level is still less than that experienced by people on many amusement park rides.

"Coming up on SRB burnout and separation. . . . Five . . . four . . . three . . . two . . . one. . . ."

The acceleration drops sharply as the SRBs use up all of their solid rocket propellant and their thrust decays to zero. This reduces the thrust driving the *Enterprise* toward orbit and also reduces the acceleration again.

We hear a couple of muffled bangs. The explosive bolts holding the SRBs to the ET have blown, severing the attachments between the three units and releasing the SRBs. We can't see or hear them, but small solid propellant separation rockets on each SRB now ignite, pushing the SRBs away from the still-climbing *Enterprise*.

The used SRBs will fall into the Atlantic Ocean below. Parachutes will deploy from their noses to slow them down before they splash into the water. Ocean-going tugboats are waiting for them and will tow them back to Cape Canaveral, where they will be cleaned, disassembled, loaded with fresh solid propellant, and made ready for another Space Shuttle launch a couple of months from now.

The *Enterprise* is now 30 miles up, traveling at more than 3,000 miles per hour. The three SSMEs in the tail of the ship are drawing their liquid propellants from the ET and driving the *Enterprise* higher and faster. Acceleration has dropped to a pleasant 1 g, but it begins to increase slowly as the *Enterprise* expends its liquid propellants and becomes lighter. The automatic pilot throttles back the SSMEs to about 75 percent of their maximum thrust to keep the acceleration within the 3-g limits.

The sky outside the porthole is now jet black, and it's possible to see the curve of the earth. The *Enterprise* is almost completely on its back, accelerating almost horizontally with respect to the surface below. The flight has been very smooth. The liftoff vibration is gone. No sound comes from the SSMEs in the tail, only a minor vibration that we can barely feel through our seat backs. We are above the earth's atmosphere, and no sound can travel in the vacuum that now surrounds the *Enterprise*.

"Mission Control, this is *Enterprise.* Coming up on SSME shutdown and ET separation. We'll take it on my count. All hands, stand by for free falling! Five . . . four . . . three . . . two . . . one . . . shutdown!"

Two shudders go through the *Enterprise.* Almost all of the propellants in the ET have been consumed. Valves in the seventeen-inch propellant pipes feeding the SSMEs close. Deprived of fuel, the SSMEs cease thrusting. We get the sudden feeling that the bottom has dropped out, the same feeling one gets coming over the first hump of a roller coaster.

Weightlessness. Zero g. Free fall.

Yes, we are now *falling* around the earth. The *Enterprise* and everything inside it is literally falling back toward the earth's surface below. The earth, being spherical, is curving away from the *Enterprise* at the same rate the ship is falling. So it continues to fall endlessly around the earth.

Almost. We're a couple hundred miles per hour short of orbital velocity. Edwards and McQuire will trim, or adjust, the orbit with the Orbital Maneuvering System (OMS) rockets once the ET has been separated.

A couple more muffled bangs are heard. The ET has separated from the *Enterprise*, explosive bolts disengaging joints fore and aft. Quick-disconnect fittings separate the propellant pipes between the *Enterprise* and the ET.

The *Enterprise* is now 115 miles above the earth and traveling at 17,400 miles per hour. This is about 200 miles per hour less than the speed required to maintain an orbital path around the earth at this height. The separated ET will therefore fall back into the earth's atmosphere and splash down in the middle of the Indian Ocean. It is the only part of the *Enterprise* that is used up, but fortunately, being just a big aluminum tank with a few pipes and valves, it is not a highly sophisticated or costly part.

Some Space Shuttles have carried their ETs into orbit where a cluster of them now forms the backbone of a rudimentary space station in a different orbit than the one we're in. The orbited ETs are vented, allowing the residual liquid hydrogen and liquid oxygen to boil off into space. Then they are cleaned, attached to the cluster of ETs, and pressurized with air. This has been a very inexpensive way to make a space base.

But the *Enterprise* now needs about 200 miles per hour additional speed to keep from following the ET into the Indian Ocean. The two small rocket engines of the Orbital Maneuvering System (OMS) in the tail of the *Enterprise* will provide that speed.

We hear Edwards's voice in our earphones.

"Houston and all hands, this is *Enterprise.* Stand by for OMS ignition and orbit insertion." The two pilots now carefully check the guidance and control computer and the autopilot, making certain that the *Enterprise* is pointed in the proper direction to make the final orbit insertion burn. "Stand by. Guidance system tracking. Platform alignment looks good. Computer instructions are in. Automatic sequencer is counting down . . . five . . . four . . . three . . . two . . . one . . . ignition!"

Each OMS rocket chamber produces 6,000 pounds of thrust for a total of 12,000 pounds. The OMS burns a mixture of monomethylhydrazine as fuel and nitrogen tetroxide as oxidizer. This combination of chemicals does not need an ignition system because the two liquids ignite spontaneously upon contact with each other; they are called a hypergolic propellant combination.

Since the *Enterprise* now weighs 215,000 pounds, we can hardly feel 12,000 pounds of thrust. Compared to the 3 g's we experienced during launch, the 0.058 g produced by the OMS burn barely pushes us back in our seats.

The orbital insertion burn of the OMS lasts for several minutes until the guidance system determines that an orbital velocity of 17,600 miles per hour has been attained. Then:

"OMS shutdown. *Enterprise* is in orbit!"

4 Living in Space

We're in orbit, falling endlessly around the earth in an easterly direction about 115 miles above the surface. We feel no gravity. Our first inclination is to unstrap and get the complete feeling of weightlessness, but a voice in our earphones stops us.

"Okay, mid-deck crew, this is Mary Howard. Welcome to orbit. Don't unstrap until I get down there to help you. Edwards and McQuire may have to adjust the orbit slightly, and I wouldn't want you to get slammed around if they do, even though the acceleration is puny. We also have to check the orbit, make sure everything's working right, and get a final 'go' from Houston to continue the mission. We could land at the Cape on the next pass, after going once around, if anything is wrong. So just stay put until I get down there."

We sigh and groan, but we sit and wait. In our earphones, we hear the pilots talking to Mission Control in Houston. Mostly, the voices repeat strings of numbers for the pilots to program into the on-board computer; these are called pad updates. Houston informs Edwards that our orbit is 120 miles by 115 miles, within a half-mile of what was expected.

During the mission, we will never be out of communication with Houston, even though we may be on the other side of the world and do not have a continuous chain of ground stations below our orbital track as did earlier manned space programs. Most of the old global tracking network has been dismantled not only to save money but because the Space Shuttles can fly into almost any near-earth orbit; it would be prohibitively expensive and politically impossible to cover the entire earth with ground tracking stations to watch each Shuttle flight.

Therefore, space itself is used. When we are out of range and over the horizon from the few existing tracking stations on the ground, our radios send their messages outward to various communications relay satellites (comsats) orbiting 22,500 miles above the earth. These relay satellites can be reached by our radio transmitters and receivers at any point in our orbit. The comsats relay radio transmissions from the *Enterprise* to existing ground stations, and vice versa.

Besides, we are in no real jeopardy if there should be an emergency. Even though we are space-borne, the *Enterprise* is still tied to earth by gravity that keeps us swinging in orbit around the planet. In an emergency, Edwards or McQuire—or Mary Howard or Khambrel Johnson, for that matter—could assume manual control, activate the retro-rockets to slow down our orbital speed, and allow the *Enterprise* to make a computer-controlled entry into the atmosphere to land on any airport runway capable of landing commercial or military jet aircraft. If something happened so that the *Enterprise* was unable to retro-fire out of orbit, we have enough air, food, and water aboard for at least seven days; in that time, NASA would prepare and launch the *Constitution* as a rescue vehicle. There are several space suits aboard the *Enterprise* as well as a number of rescue balls, inflatable three-foot-diameter plastic bubbles that will hold one person and provide enough air for several hours so that people could be transferred from one Shuttle to another through the vacuum of space. All possible emergency situations have been studied and have solutions ready to be put into action.

My thoughts about emergencies are interrupted as Mary Howard "swims" down through the opening from the flight deck. She has removed her Snoopy helmet, and her long curls float around her head in zero g. "Okay, troops, one at a time. Dr. Cleveland, you first." She floats across the mid-deck to the right-hand seat, helps the scientist unstrap, and disconnects her Snoopy helmet.

Then she floats over to me. She disconnects the communications wires running to my helmet, and I make the mistake of taking it off. It begins to float away.

"Grab it and put it in your personal equipment locker before you lose it," Mary warns. "In fact, don't let go of anything in zero g unless you're parking it for immediate reuse or sending it to someone across the deck. If you just turn loose of something and forget it, you'd be surprised how quickly it gets lost and where it finally ends up."

Then I unstrap by pushing the single button in the middle of the chest clasp that holds all the restraint belts. The belts float away from me like big, wide, flat snakes. I float away from my seat cushions. Mary moves over to help you.

Weightlessness is a sensation we've experienced for only a few seconds at a time in training flights aboard special jet planes. But here, weightlessness is the natural way of life, and it's continuous. At first, it seems like an endlessly falling sensation. But that is quickly dispelled because our eyes tell us that the cabin and the rest of the *Enterprise* are falling with us, and this tricks our minds and eliminates the falling sensation. Weightlessness very quickly becomes natural. Weight is felt only when one of the *Enterprise*'s rocket motors is operating, providing acceleration.

Weightlessness also makes the mid-deck seem much larger because we're no longer confined to standing on the floor. We can float in any attitude, or orientation, anywhere in the compartment, even on the ceiling like a human fly.

We laugh as we flounder about in the air, and we

laugh even harder when we look at one another's faces. With gravity no longer pulling downward on our facial features, our cheeks appear bloated and our eyes puffy. This facial distortion is another consequence of weightlessness. Remember the old photographs of the astronauts and cosmonauts in orbit during the *Apollo-Soyuz* international mission in 1975? Remember how different the men looked in comparison to photographs of them taken on earth?

Getting about the cabin in weightlessness takes a bit of practice. One of the first things we learn, to our embarrassment, is not to float out of reach of a handhold. If you happen to float into the middle of the deck where you can't reach out and touch something with your hands or feet, you'll just float there until someone comes to get you or until you are wafted to a wall by air currents. And although it is possible literally to fly through the air from place to place in the cabin, you can't be too energetic about it, for the faster and harder you push off, the harder it is to stop when you reach your destination. Although weight has disappeared, the other characteristics of matter remain—mass and momentum.

Our first order of business is to stow the seats in which we rode during liftoff. We give Mary a hand at this. Each seat disconnects from its fittings on the floor and folds like a beach chair or camp stool by releasing latches and removing retaining pins from folding joints. The weight of the seats doesn't hamper Mary a bit because they weigh nothing now. And she's had experience with them.

"Be careful," she warns us. "Don't catch your fingers when we fold them!"

The three seats, folded flat, are stored in lockers around the walls of the mid-deck. With the seats out of the way, there is even more room in which to move around.

I maneuver over to the little porthole through which you were able to look during liftoff. Sure enough. There we are, in orbit on our back. The *Enterprise* is riding in orbit with its belly up, or away from the earth below, and its cargo bay toward the surface. It really doesn't make much difference at this point. The *Enterprise* can fly in orbit in any attitude and has control rocket thrusters to permit Edwards, McQuire, or Johnson to change the attitude as desired.

The porthole offers a limited view, but it's possible to see the curve of the earth at the horizon. The outline of the globe is not sharply defined; I can see that the earth has a definite atmosphere because the edge of the planet is fuzzy against the dead black background of space. The surface is nearly covered with clouds, making it very white and very bright. Through scattered holes in the clouds, I can see a flat blue surface that must be ocean. Occasionally, a brown-green patch appears that must be an island. There is no doubt about it: Earth is a cloudy water world.

I regret the view is so restricted. But, after all, the *Enterprise* is not a tourist ship. It's a space truck. Perhaps someday in the near future, though, there will be spaceships with windows like a jet airliner's so all the passengers can see out.

Mary gives the three of us a refresher course about all the mid-deck life-support equipment, personal hygiene systems, and food preparation gear. We've had several weeks' training in the simulators at Houston, but this is for real, and Mary wants to make sure we haven't forgotten anything. Living in space is different from living on earth in some ways—and the same in others. In both cases, you have to remember certain procedures.

There is a zero-g toilet on the left side of the rear bulkhead. It's inside a compartment with a door that closes for privacy, and it looks almost normal and ordinary except for the seat belt and the urine receptacle in front. It operates by applying a slight vacuum to draw human wastes into it. Then it seals itself with a sliding lip that slips under the regular seat. The seat belt and two handholds are provided to insure that you don't float away from the seat. The urine receptacle is like the old relief tubes in military airplanes and is similar to the unit aboard the Skylab. The toilet compartment is well-ventilated to remove odors and is well-equipped with extra-absorbent wiping towels just in case things happen to get out of control.

The personal hygiene center is just forward of the toilet compartment and can be separated from the rest of the mid-deck by a curtain for privacy. There is no running water. In weightlessness, running water would go all over the place, forming big and little globs and causing all kinds of problems. We can wash up by using moistened towelettes. Those who must shave may use either their ordinary electric razors or blade razors with shaving cream. Brushing one's teeth is different because we must swallow the toothpaste instead of spitting it out; after all, there's no place to spit in weightlessness. The toothpaste was developed for the *Apollo* program, and we can swallow it without harm.

Each of us has several personal hygiene lockers for our own color-coded towelettes, toiletries, and other belongings.

Since we are now in orbit and weightless, it is possible to open the personal lockers on the forward bulkhead without having the contents spill out. But they may float away, so we still have to be careful. In our lockers are changes of clothing plus those personal items each of us has put aboard. I have a spare pair of eyeglasses in case something happens to the pair in my pocket. And I have those mementos for myself plus a couple that I'll give to certain special friends after the flight.

Our personal hygiene is our own responsibility. In the early days of space flight, everything was done by the numbers according to a completely detailed flight or mission plan. This was even true to some extent in the Skylab program. And facilities for living in space were sometimes very primitive, with limited capability of carrying personal gear. The situation even led to jokes, like the one where Mission Control sends up the order, "Okay, fellows, today you are scheduled to change underwear. Joe, change yours with Jim. Jim, change yours with Bill. Bill, give yours to Joe." Conditions weren't really that bad, but some of the older astronauts at Houston spin space tales that sound pretty grim. The Shuttles offer high living in comparison.

Sleeping in orbit will be a real experience to talk about. Mary explains the arrangement again to refresh our memories.

"We have three bunks over here on the right side of the cabin. There is one location for a Skylab sleeping bag to be hung on the aft wall of the bunks, but nobody likes to use it. We'll be in orbit for only twenty-four hours, more or less, but everyone is scheduled to get at least eight hours' sleep. Since we have only three sleeping places aboard for seven people, we'll follow the old Navy 'hot bunk' system that was used on destroyers and submarines. When you finish your scheduled sleep period, there'll probably be someone else ready to occupy the sleeping space with his or her own sleeping bag. Now, you all know the work schedule and how important it is to stick to it. So let's not have anybody sleeping late, because somebody else is waiting to crawl in when you're scheduled to wake up."

Each of us has our own sleeping bag, which is stowed in a locker when not in use. It might be fun to sleep out in the cabin in weightlessness, but the cabin will always be occupied. Besides, a person might float around and bump into something while asleep. The sleeping bags are lightweight beta cloth, just enough to keep us from floating away while asleep. They don't need to be warm because the cabin temperature is always controlled for comfort.

Mary unstows the wardroom table where three of us at a time can take our meals. Typical of the internal design of the *Enterprise*, the table is cleverly engineered. It slides out of a slot in the forward bulkhead between two rows of lockers. It has two legs at each outside corner that do not support it but serve to steady it. Three stools swing out from the legs, each with a lap belt attached to it so a person can be secured in place while eating. The table can also be used for working space if someone needs to make notes, do calculations, or work on a piece of gear.

The galley is on the left-hand cabin wall forward of the personal hygiene center. It, too, is an engineering miracle of compact, utilitarian design. Our food can be heated here in a microwave oven. There is a water supply for use with freeze-dried and dehydrated foods, and waste compartments are provided for storage of used dishes and utensils.

Some of the food aboard the *Enterprise* is exactly the same as that eaten by the *Apollo* astronauts, some is the same as that eaten by the Skylab crews, and some is the result of more recent NASA development in food storage and preparation. Many of the containers look familiar because in our everyday shopping, we've grown accustomed to food items and packaging originally developed in the space program—such as pull-top cans, pop-top cans, and freeze-dried foods.

Mary floats to the hatchway between the mid-deck and the flight deck and calls, "Okay, Dave, tell Dick we're secured for orbital operations on the mid-deck." The copilot replies, but we can't hear him, and Mary tells him, "As ready as we'll ever be."

To our surprise, both pilots float through the hatch from the flight deck. Edwards tells us, "We're about to have a little ceremony here. Won't take long because we've got work to do, but as commander of the *Enterprise*, it's my duty to follow naval, aeronautical, and astronautical tradition by carrying out this ceremony for each new groundhog who travels into orbit for the first time. Now, will you three mid-deck people sort of line up against the forward bulkhead, please—and orient yourself with your feet toward the floor, as I'm doing? By the way, when in weightless condition, one always assumes the attitude and orientation of the commander of the vessel when talking to him. Common space-going courtesy of command. I know it's difficult to stand, but sort of float, if you will." Edwards is conducting this ceremony with mock seriousness, working hard to suppress a grin.

We are then inducted officially into an unofficial but very traditional fraternity. It began in the *Mercury* program, the United States' first manned space flight project. It continued quietly through the *Gemini* program, the *Apollo* flights, and the Skylab missions. Its roots go back to the days when ocean-going ships crossed the earth's equator and lowly "pollywogs," or first-timers, were crudely initiated into the exhalted ranks of those who had previously crossed the equator, the lordly "shellbacks." The specific secret society into which we're inducted began among pilots during World War II. It involves a question to which a very precise answer must be given under all conditions. Naturally, since it's a secret society of those who have participated in space travel, I'm not at liberty to divulge the nature of the secret ceremony. Suffice to say that the three of us from mid-deck are proudly inducted into the ancient and honorable Order of Turtles.

The ceremony takes only a moment. Mission Control probably knows about it because it happens on every flight where "groundhogs" go into space for the first time. But nothing ever appears in the recorded mission tapes to indicate it takes place. So it *officially* doesn't happen. It might cause problems for NASA if it were known, which is why some high NASA officials stay ignorant of it and others are wise enough to look the other way.

Sporting our little gold turtle pins on our coveralls, Dr. Cleveland and you and I are invited up to the flight deck for our first real view of outer space.

5 Working in Space

We three new Turtles float up through the hatch from the mid-deck to the flight deck, while Edwards, McQuire, and Mary Howard stay below to eat lunch. The flight deck would have been crowded had the three of them been there, too. Khambrel Johnson is on duty on the flight deck, standing before the on-orbit controls of the *Enterprise* located in the aft bulkhead.

Khambrel is able to stand, rather than float, because the soles of his flight shoes and the floor in front of the control panels are covered with velcro. Velcro consists of two types of material, one with miniature fiber hooks and the other with a soft wooly surface. When the two dissimilar types of velcro are brought into contact with one another, they stick together. They can easily be separated with a ripping sound that's often difficult to get used to. The velcro keeps Khambrel's feet on the deck and allows him to work in weightlessness.

The view from the flight deck is tremendous!

We can see forward through the pilots' windshields. And above Khambrel's head are two more windows used for payload handling, rendezvous maneuvers, and other orbital operations.

The *Enterprise* is still upside-down, its belly toward space. The sun is behind us, and we are coming up on the sunset line over the Indian Ocean. The sky is absolutely black, and no stars are visible because the glare of the sunlit earth below is far too bright. Actually, as we float with our feet floorward, the earth is over our heads. But, strangely, it doesn't make any real difference to us.

Khambrel tells us, "I'm going to open the cargo bay doors now to deploy the coolant radiators. You'll be able to see the sunset through the windows here in front of me."

He manipulates controls. Suddenly a crack of light appears through the two windows facing aft into the cargo bay. The huge cargo bay doors, each sixty feet long, begin to swing out, opening the bay to space. Along the inside surface of the forward part of each door is a curved coolant radiator that must be exposed to space during orbital operations to carry off the excess heat generated by the electrical and life-support systems of the *Enterprise*. Many writers have told of the "bleak cold of space," but that isn't so. Space is a very good vacuum by earthbound standards, and it is neither hot nor cold. The coolant radiators on the inner side of the cargo bay doors get rid of heat the only way it can be done in space—by radiation. They radiate heat out into space just as an electric heater radiates heat in your home on a cold winter day. Only the *Enterprise*'s coolant radiators don't glow red-hot the way an electric heater does.

Now that the cargo bay doors are open, we can see aft over the bay to the tall rudder and bulging OMS enclosures on the tail of the *Enterprise*.

The sun is glaring through the windows, bright and white as it approaches the curve of the earth behind us. It does so with amazing swiftness, much faster than any sunset on earth. Suddenly the lower edge of the sun touches the wispy edge of the earth's upper atmosphere over the horizon. Immediately, color springs out on both sides of the sun's disk—blue, orange, yellow, and red. Within a minute, the sun has slipped below the rim of the world, but the band of light caused by sunlight diffused through the atmosphere remains. It glows with color that slowly fades.

And the stars come out. They become visible slowly, the brightest ones first, as our eyes accustom themselves to the absence of sunlight. They are bright, clear, unwinking—and there are millions of them.

The earth below is still strangely visible because of the moonlight falling upon it. Through the side windows, we can see a first-quarter moon. It doesn't look any closer from orbit than it did from earth, but it wouldn't take much more speed to get there—only about 10,000 miles per hour, which is little more than half the orbital velocity of the *Enterprise* right now.

So near and yet so far. People have walked the powdery surface of the moon in years gone by, and before the century is out, they'll walk it again, this time in search of raw materials for space industrialization.

"Houston, this is *Enterprise.*" Khambrel's voice breaks the silence. "Cargo bay doors are open, and coolant radiators are deployed."

"Roger, *Enterprise.* We read you loud and clear and confirm telemetry shows cargo bay doors open and coolant radiators deployed."

Mission Control can confirm our operations in space because the *Enterprise* carries radio transmitters that tell the experts in Houston how various systems aboard the orbiter are working. This is called telemetry and has been used in missile and space programs since the early days at Peenemünde, Germany, during World War II. Our radio links to Houston are through a comsat (communications satellite) 22,500 miles above us, which is linked to a ground station near Perth, Australia, which is linked by another comsat to a ground station on the West Coast of the United States, which is in turn connected through microwave and telephone lines to Houston. Space research has wrought a quiet revolution in worldwide communications, and the radio links with the *Enterprise* in orbit are only a miniscule part of the huge volume of communications traffic handled continuously by the comsats.

"Houston, this is *Enterprise.* I'm going to start the roll transition to the Landsat deployment attitude," Khambrel reports.

He is talking about the new earth resources surveying satellite, or Landsat, stowed in the *Enterprise's* cargo bay. The mission plan calls for Khambrel and Mary to deploy it into orbit over the Indian Ocean on our second orbit. Later in the flight, Khambrel will maneuver the *Enterprise* to pick up an inoperative earth resources satellite that is being replaced by the Landsat in the cargo bay.

To deploy the new Landsat, Khambrel must turn the *Enterprise* over and then eject the satellite into space with a precise separation velocity to insure that it goes into the proper orbit. There is no doubt that Khambrel is a genius at this sort of thing. His hands fly over the control panels, making adjustments. In spite of all the bioastronautics experiments, personnel selection tests, and training procedures that have evolved in the space program, behavioral scientists have discovered that it is difficult to measure a person's ability to think in four dimensions simultaneously—three spatial dimensions plus time. Khambrel has a natural talent for such thinking, which makes him an outstanding payload specialist.

"May we stay and watch?" I ask, anxious not to be underfoot, so to speak.

"Sure," Khambrel replies. "The mission schedule calls for Dick and Dave to eat right now, and Mary's busy below, so there's no reason you can't stay up here on the flight deck until we deploy the Landsat. *But,* I want all of you to stick your feet to velcro and grab a handhold. I won't roll the ship rapidly enough to throw you around, but the visual cues may disorient you unless you're hanging on tight."

I plant the soles of my shoes firmly on a velcro deck pad. This isn't as much fun as floating around in weightlessness, but it gives one a definite sensation that gravity force is holding one's feet to the floor. I grasp a handhold on the back of the pilot seat.

The attitude, or orientation, of the *Enterprise* in space is determined by a gyro-stabilized platform that maintains a constant position with reference to the stars. To change the orbiter's attitude, Khambrel must ignite the vernier OMS rocket engines located in the nose and tail. Such attitude control rocket thrusters have been used on all manned spacecraft to date. The attitude control thrusters of the *Enterprise* can be operated according to a computer program that tells the thrusters when to thrust, which thrusters to thrust, and when to stop thrusting; the computer gets its instructions from a keyboard on Khambrel's control panel and from position sensors on the gyro-stabilized platform. Or Khambrel can operate the thrusters manually and control the ship's attitude with a calibrated eyeball—that is, by using his own trained senses and judgment. Surprisingly, both methods are almost equal in precision.

At an altitude of twenty-five miles, the two Solid Rocket Boosters have used up all their rocket propellant and are separated from the *Enterprise* by small separation rockets, as the Space Shuttle Main Engines continue to thrust the spaceship toward orbit. The SRBs will land by parachute in the Atlantic Ocean, where they will be recovered for refurbishment and reuse.

About 115 miles up, the Space Shuttle Main Engines have exhausted all the liquid hydrogen and liquid oxygen in the External Tank, and it is separated from the *Enterprise*. The ET will fall back to earth in the Indian Ocean—the only primary Space Shuttle part that is not reusable.

Some External Tanks have been carried into orbit to form the backbone of a rudimentary space station. This drawing shows a Space Shuttle bringing materials to a demonstration satellite solar power station that could be placed in orbit to beam microwave energy to earth.

The final acceleration of the *Enterprise* into earth orbit is made with the Orbital Maneuvering System (OMS) rocket engines housed in the tail of the orbiter. The *Enterprise* goes into orbit on its back about 115 miles up.

In the weightlessness of orbit, a crew member floats across the mid-deck of the *Enterprise*. The ship's three bunks may be seen on the far wall, where another crew member dozes in a Skylab-style sleep sack. The zero-g toilet is in the compartment to the right. The cylindrical structure is the air lock leading to the cargo bay.

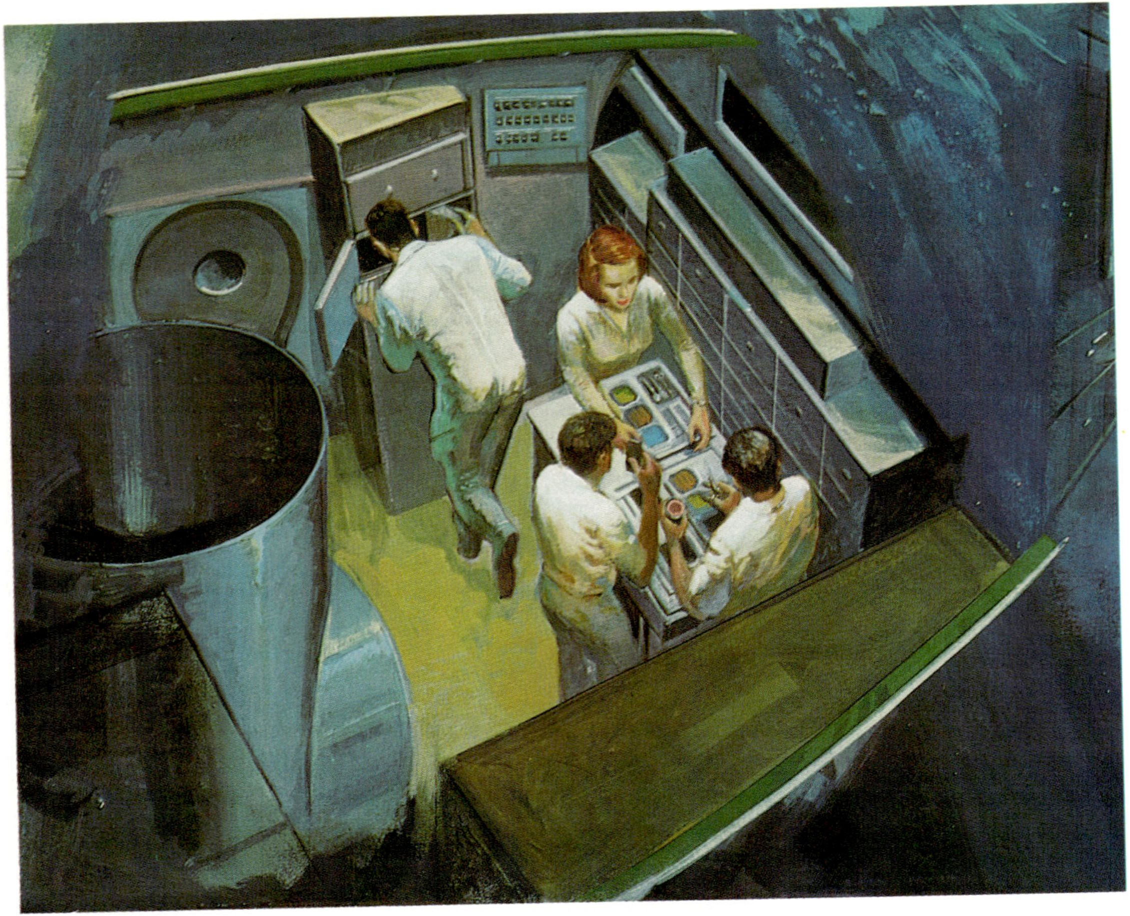

Three members of the *Enterprise* crew eat at the wardroom table on the forward bulkhead of the mid-deck, while a fourth member freshens up at the personal hygiene center.

This view from the aft windows of the flight deck shows part of the open cargo bay and the manipulator arm as it extracts the new Landsat from its cradle in the bay. Placing various types of satellites in orbit, servicing them in orbit, or returning them to earth for service are some of the main functions of Space Shuttles.

On the jointed manipulator arm, the Landsat is lifted out of the *Enterprise*'s cargo bay, and its two solar power panels are extended. Note the curved coolant radiators on the upper side of the forward cargo bay doors.

"I'm going to roll us over by computer control because I need an operational check on the computer system," Khambrel explains. "Besides, it's easier to let the computer do it. I'll save myself for the Landsat retrieval—don't worry, I can override the computer with the manual controls if something goes wrong." He consults several pages of a booklet attached to the ceiling over his controls, writes some numbers on a slip of paper on the console, and begins punching buttons on the computer input keyboard—the DISKY—on the panel. "Okay, here goes." He pushes a final button.

The computer has been instructed by Khambrel to roll the *Enterprise* very slowly so that the cargo bay doors face away from the earth. Thrusters in the nose and tail ignite in short bursts, producing 870 pounds of thrust each time they burp. There is no flame or sound. Over the nose and back by the tail, we can see brief plumes of incandescent gas as the thrusters fire.

The universe begins to revolve slowly, or at least it appears that way as the *Enterprise* begins a slow roll.

"It should take about a minute to complete a half-roll," Khambrel explains. "I'm in no hurry, and a gentle roll puts less stress on the cargo bay doors and radiators. It also saves thruster propellant. In the meantime, watch, because the sun is coming up ahead of us."

Halfway through the roll, we can see the earth's horizon ahead of us getting brighter and brighter. We watch a reverse of the sunset we saw a short time before. With startling suddenness, the brilliant white disk of the sun pops over the curved horizon. The surface below is still dark. But as the sunrise progresses, the tops of the cloud formations over the Pacific Ocean become illuminated by the sun's rays.

Another burst of the thrusters stops the roll. We are now "flying" in conventional attitude with the big black belly of the *Enterprise* pointed toward the earth's surface.

Khambrel has maintained a steady dialogue with Houston throughout the maneuver, reporting and confirming activities. In between, he remarks to us, "We'll be coming up on the West Coast of the United States shortly. In this attitude, you can't see too much of the surface below, but you may be able to pick up a view of San Diego slightly to the right of the nose. We'll be tracking across the southern states and heading out over the Gulf of Mexico west of the Florida peninsula."

Mary's head appears in the hatch leading to the mid-deck. "Okay, Dr. Cleveland, you're scheduled to eat now. Your lunch is in the galley warming oven. I have to get up here to help Khambrel."

There is an exchange of people on the flight deck, but we are permitted to stay on to watch the deployment of the Landsat from the cargo bay. It takes two people to handle this activity, so Khambrel and Mary are both busy at the control consoles on the rear bulkhead of the flight deck. Dick Edwards sticks his head up through the mid-deck hatch to keep an eye on things. As commander of the *Enter-*

prise, he is essentially the captain of the ship and responsible for everything that happens.

"No, no, stay on the flight deck," Edwards tells us with a wave of his hand. "I can monitor things from here just as well."

Khambrel begins checking out the manipulator arm, a fifty-foot jointed mechanism that is stowed along the left side of the cargo bay. It is very much like a human arm, with a shoulder joint at the left forward corner of the cargo bay, an elbow joint about halfway out, a wrist joint, and a manipulator claw on the end. Like the human arm, all joints not only bend but rotate as well. This ingenious device was designed and built in Toronto, Canada, emphasizing the true international nature of the Space Shuttle program. As we pass over the southern United States, Khambrel checks the manipulator arm by exercising all its joints and motions.

Some Space Shuttle missions carry two manipulator arms, one on each side of the cargo bay, to assist in deploying or retrieving satellites or to help in assembling the large structures now being built in space. These manipulator arms played an important part in helping to construct the Space Base and the Space Industrial Park, which are in a different orbit than ours, so we won't see them on this flight.

Mary operates the controls that power-up the Landsat in the cargo bay, while Khambrel checks the readings of instruments reporting on the condition of the clamps and latches holding the Landsat in its cargo bay cradle. This is a joint effort, a step-by-step process of checking out, cross-checking, and double-checking. The Landsat in the cargo bay cost many millions of dollars to build and will provide information worth many millions of dollars more than it cost. It must be handled properly and checked out carefully so that it will operate as planned when separated from the *Enterprise* and placed in a free-flying orbit of its own. This procedure is far superior to the launches of a decade ago when satellites had to be launched aboard expendable rocket boosters, sometimes to fail when placed in orbit. Satellites can now be built more cheaply because they can be checked out by human beings in orbit before being left on their own. And if they don't work as planned, they can be brought back and fixed. This saves a lot of money, materials, and energy.

Khambrel first attaches the claw of the manipulator arm to a deployment fixture built into the Landsat structure for that purpose. His remote readouts tell him that the arm is attached and the claw latched. Mary then activates controls that remove the electrical plugs connecting the Landsat to the electrical power supplies of the *Enterprise.* Another control releases the Landsat from its cargo bay cradle.

"It's free and ready to be deployed," Mary reports, scanning the telemetry readouts that tell her the Landsat is successfully operating on its own auxiliary batteries and nuclear power supply.

"Okay, coming out of the bay now," Khambrel murmurs, intent upon the careful operation of the manipulator arm.

Over their shoulders, we can see through the aft windows into the cargo bay. Slowly, as Khambrel operates the arm, the boxy Landsat is withdrawn from the bay. Finally, extended on the end of the arm, it is clear of the *Enterprise* but still grasped by the claw.

"Solar panels extending," Mary reports.

Like a butterfly emerging from its cocoon and extending its blue iridescent wings for the first time, the Landsat slowly unfolds its two panels covered with blue solar cells that will convert sunlight into electrical power for the satellite's electronic equipment. The panels unfold once, then unfold again. Against the black background of space, it is difficult to judge the Landsat's true size. But it now looks so big with its solar panels deployed that one wonders how it ever managed to fit into the cargo bay in the first place.

"Solar power is now on the satellite's main electrical bus," Mary reports, her conversation being heard by Mission Control as well as those of us on the flight deck. "Attitude control system coming into alignment. The star tracker has just acquired Canopus! Okay, Khambrel, she's stable and will hold her attitude when you release her. All her systems appear to be functioning."

Mission Control confirms that their radio telemetry indicates all Landsat equipment is operating properly.

The sun goes down behind us for the second time today. The Landsat takes on an eerie aspect as it follows the *Enterprise* on the end of the extended manipulator arm.

"*Enterprise,* this is Houston. Stand by for apogee

on my mark in about thirty seconds," the voice from Mission Control advises.

Khambrel will release the Landsat at the apogee of the *Enterprise*'s orbit. The satellite will then follow the *Enterprise* until Khambrel changes orbit slightly with the thrusters to get well away from it.

Mission Control marks the countdown, "Four . . . three . . . two . . . one . . . release!"

Khambrel releases the latch on the claw at the end of the manipulator arm, and the arm withdraws from the Landsat, leaving the satellite suspended on its own in orbit.

"Hang on, everybody," Khambrel warns. "I'm going to move us gently away from it now." He fires thrusters manually, and the Landsat appears to move away from us. "Okay, now I'll translate away from it and reestablish our orbit. Tell Dr. Cleveland that she can come up here and start taking data with her experiment on our next pass over the Pacific."

Mary secures her control panels and turns to us. "Time for you folks to eat. Got to stick to the schedule, you know. You're to start your experiment in about three hours."

Down on the mid-deck, the two pilots are tucked into corners of the cabin, reading. Dr. Cleveland goes up to the flight deck to begin running her experiment. And we eat lunch.

Eating in space is no problem except that everything floats away if you let go of it. Some foods are packaged in small pop-top cans that are electrically warmed in the galley oven. Others are freeze-dried just like many of the things you can buy in the supermarket at home. The freeze-dried foods are in flexible plastic bags with nipples through which water can be added; after the bags are shaken vigorously to mix the water and other contents, diners suck on them like a baby bottle while squeezing the sides to force the contents out. Most solid foods are of the sticky type that hold together, like mashed potatoes, because crumbs are messy and go all over the place. We eat with regular utensils—plastic forks and spoons.

Swallowing without gravity is no problem because gravity isn't really needed to get food from your mouth down your esophagus to your stomach. After all, you can eat a cracker while standing on your head with your whole system working *against* gravity on earth!

I try an experiment that I've always wanted to do. Carefully, I squeeze some water out of a plastic bag into a clear plastic cup, and, by swinging the cup with my arm, I get all the water to settle into the bottom. Then I let it float free in the cabin to see what happens.

The water in the bottom of the cup does not float out in a glob. Its surface tension, or wetting ability, keeps it attached to the cup bottom. But, as we watch, that same surface tension causes a film of water to creep up the sides of the cup, over the edge, and down the outside until the entire cup is covered with a layer of water. It's a successful experiment demonstrating the positive wetting characteristics of water.

During our meal, we have occasionally felt a

gentle nudge as Khambrel fired the thrusters to adjust our orbit. He is not only moving away from the new Landsat but changing our orbit slightly so we can rendezvous with the old Landsat that has stopped working. In about twelve hours, Khambrel will have the *Enterprise* positioned so that he can reach out with the manipulator arm, grasp the ailing Landsat, put it into the cargo bay, and secure it in the cradle to be taken back to earth for repairs.

Dr. Cleveland comes through the hatch from the flight deck accompanied by Khambrel, who's now scheduled for lunch. The doctor looks happy.

"The equipment works beautifully!" she explains. "I had to make a few unplanned adjustments and change a few details, but I'm getting fantastically good data! Not only can I sense wave heights and wave separations, but I can even discriminate between long-duration ocean swells and the sort of short chop that's generated by nearby storms. I'm certain I'll be able to find some correlation between wave action and ocean weather patterns. It's going to take a lot of time to get the data reduced, but this will certainly help us forecast the weather and perhaps even shed some light on changes in climate."

We congratulate her on the success of her experiment and then move up to the flight deck and join Mary. Dave McQuire follows us, slipping into the copilot's seat to help monitor the systems aboard the *Enterprise* during our experiment, which will place a heavy load on the electrical supplies and the cooling system.

Our experiment in space processing is relatively simple. We want to see if we can make a very lightweight, high-strength plastic material. We're going to start with simple polystyrene, the type of thermoplastic usually used to make hobby model kits and toys because it is cheap and because it melts at a low temperature and can be remelted. To make it very lightweight, we're going to whip it to a froth in its liquid state, mix dry nitrogen gas with it, and make it into a foam. Nitrogen will not react chemically with any of the materials in the experiment.

There are many types of foamed plastics already in use, but they all suffer from the same problem: When you foam plastic, it loses strength. The more it is foamed, the weaker it becomes.

This problem is compounded by the difficulty of obtaining, in earth's gravity, a uniform mixture of bubbles and plastic throughout the foamed material. On earth, gravity causes the bubbles to rise through the plastic, just as air bubbles rise to the top of water run into a glass from the kitchen tap. In weightlessness, where there is no "up," we should be able to get a uniform mixture of plastic and nitrogen bubbles so that the resulting foamed plastic has the same density and strength throughout.

To increase the strength of our foamed plastic, we're going to borrow a trick from Mother Nature. She makes an equivalent plant material of cellulose with open voids; tough fibers of cellulose run through it for strength. We call this strong, light material wood.

But producing wood means cutting down forests of trees. And, although we have lots of forests and trees, the demand for wood keeps increasing all the time. We think that we can make a material that will be comparable to wood but lighter and stronger by adding long metallic filaments, called whiskers, to the foamed plastic and making these whiskers line up by applying a magnetic field to the material before the plastic cools and hardens. When we tried this in our laboratory back on earth, the metallic whiskers began settling to the bottom of the plastic before the plastic could harden, and we couldn't make a material with equal strength throughout. In the weightlessness of space, though, we should be able to get a uniform mixture of plastic, bubbles, and whiskers.

It took a long time to design and build our experiment because it has to be run in the cargo bay of the *Enterprise* by remote control. There wasn't room for it in the pressurized mid-deck, where we could have worked on it directly.

We obtained our metallic whiskers from a research scientist who made them in space aboard a previous Shuttle flight. It isn't possible to make long metallic whiskers on earth because gravity breaks them off before they get long enough to be of much use.

Our polystyrene supply consists of six blocks of clear plastic, each in its own container, so that we can make several samples with different characteristics. We will heat each sample to 250 degrees Fahrenheit to melt it. Then we'll inject dry nitrogen gas into the sample and whip it with a blender-type device you developed. (I recall how you invented it by watching an ordinary food blender whip up a milk shake in the kitchen.) While the foamed plastic is still in its molten state, the metallic whiskers will be dumped into the mix. A magnetic field will be applied to line up the whiskers like little bar magnets. Then the sample will be cooled to harden it and automatically moved to a storage container so that the next sample can be processed.

The frustrating element in the whole experiment is the fact that we won't be able to check the samples immediately. We can monitor remotely all the factors of the process as they are taking place, and control times, temperature, blender speeds, volume of gas, amount of whiskers, and other elements. But we can't examine the results until we get back to our earthbound laboratory where we can look at the samples under microscopes and subject each of them to numerous tests.

Such experimentation cannot truly be considered scientific. It could be called empirical development because we are trying to obtain a definite end product by a cut-and-try method as we go along. There is no way to conduct the experiment on earth, and it can't be done in an unmanned satellite. We have to be on hand to run it, even by remote control, and we have to bring it back to earth intact so we can study the results.

Our hands shake a little with anxiety as we run

the first sample through the experiment, but we gain confidence by the time we have processed the third sample and see that we are getting good results. The procedure becomes simple. Move plastic sample into oven. Oven on. Raise plastic temperature to melt point. Blender on. Inject nitrogen gas. Keep careful watch on blender speed and torque to insure a good mixture. Add whiskers. Continue blending while applying magnetic field. Shut down blender, and withdraw blender blades from the mix. Shut off oven. Monitor mixture temperature until it has cooled below the hardening point. Activate ejector to force this sample from the oven, and inject a fresh plastic sample. Then do it all over again.

I'm glad that you are taking copious notes as we proceed, and are also tape recording a running verbal commentary. The notes and commentary are part of our total data package. They may be as important as the samples themselves once we get back to the lab. If our experiment fails, it will be more than nine months before we can obtain space assignments on another Shuttle flight, and it will put a large hole in our budget to have to fly again.

Four hours later, our work is done. We have six samples of space wood stowed in our equipment in the cargo bay. Our every urge is to get a look at those samples, but that will have to wait until we land.

In a year, the new Space Industrial Park will be completed, and we'll be able to move into an experimental module there for a month or more if we proceed with the development of our space wood. The Space Industrial Park is a cluster of External Tanks and other enclosures placed in an equatorial orbit just below the Van Allen radiation belts—high enough to be totally out of the earth's atmosphere for true space conditions, yet below the harmful radiation of the belts. We may be able to set up a pilot plant to make enough space wood so that it can be used for practical purposes on earth and in space. Someday there may even be a big space factory making our material.

Our proposed product is only one of thousands that can be made only in space or made better in space. We are learning, too, that we can make in space almost anything that we can make on earth. So the long-range dream is to remove heavily polluting industries and factories from earth and relocate them in space where they cannot pollute anything.

Now it's time for supper and then for some sleep. I'm tired, and you look bushed, too. We don't mind calling it a day, especially since tomorrow promises the spectacle of retrieving the ailing Landsat from orbit and then landing on earth.

6 Trouble and Homecoming

Somebody's shaking me out of a very sound, dreamless sleep. For a split second, I tense because I seem to be falling endlessly. Then a slight pressure from my sleeping bag reminds me—I'm in orbit, weightless. And you are the one who's shaking me awake. It's time to get up.

Mary already is out of the top bunk, and Khambrel is folding up the sleep sack on the wall.

Sleeping in space was not the unusual experience I thought it would be. I drew the bottom bunk assignment, which meant that I slept on the bottom of the tier of three bunks, but upside down, or back-to-back with you in the middle bunk. Not that it seems to make any difference in the weightlessness of space, of course, but it did seem strange when I climbed into my sleep sack eight hours ago.

I turn on the bunk light, draw the curtain closed, and take the opportunity to change into fresh clothes. It is easy to change clothes in the cramped space of the bunk because I float, so the clothes slide on and off very readily. My turn comes to use the personal hygiene facility—that name always makes me grin. I freshen up with a moist towelette, brush my teeth, and swallow the toothpaste. I have to dampen my hair slightly to get it to stay in place. In weightlessness, hair has a tendency to float all over the place, making it a nuisance, which is perhaps why the Shuttle flight crews sport the super-short "buzz cut."

You and I are experienced space travelers now, so we cook our own breakfast in the galley. It's really pretty good, even if it isn't the sort of fare we are used to at home. The orange juice powder becomes palatable if mixed at double its recommended strength. The scrambled eggs in the pop-top can aren't exactly great, but they're tasty. The sausage from another pop-top can is delicious. We each have a soft breakfast roll, one that doesn't produce crumbs, and milk in a squeeze bottle.

When we finish the meal, there is no need to wash the dishes. The used eating equipment is simply stowed away in a garbage locker just as they do it aboard a jet airliner.

We have heard rumors that the food isn't quite as good on the thirty-day flights where all of it must be freeze-dried or otherwise preserved. Since ours is only a twenty-four-hour flight, our food could be put aboard like airline food, reasonably fresh and, thanks to refrigeration, almost as good as most restaurant fare on earth.

Dr. Cleveland is bubbling over with excitement. Her laser wave experiment has given her so much good data that she says she has trouble believing it. "I could never have done this sort of thing in an unmanned satellite. The experiment would have failed because I couldn't have made the on-the-spot adjustments I discovered were necessary."

You tell her we're happy with the way our experiment went, too, though we'll have to wait until we get to the lab to find out just how successful it was.

Mary's head appears through the flight deck hatch. We are now so accustomed to weightlessness that it doesn't seem unusual to see her floating upside down with respect to us.

"Do two of you want to watch Khambrel recover the Landsat?" she asks.

Dr. Cleveland turns in for some rest; she's been up for almost twelve hours, running her experiment. Edwards and McQuire are also resting in two of the bunks, Edwards with his Snoopy helmet on so that he can hear the radio communications and the intercom. As spacecraft commander, he's never far from the action, even when he's not performing it himself.

So the two of us go up to the flight deck again to watch Khambrel work.

Through the overhead rendezvous windows, we

can see the old Landsat hanging in space next to the *Enterprise.* It is difficult to judge its distance from us because there are no points of reference out there for comparison. The Landsat looks very much like the one Khambrel put into orbit yesterday, but this one is an earlier model. Something has gone wrong with its image storage system, and it can report to ground stations only the view that can be seen from its cameras at the time a ground station requests a data transmission. It must be brought back to earth for repairs and then returned to orbit in another Shuttle flight.

Khambrel is maneuvering the *Enterprise* with the thrusters. We feel their gentle shove as they operate. We must be getting accustomed to weightlessness now because those "gentle" shoves seem much stronger today.

"Houston, this is *Enterprise.* I'm ready to reach for the Landsat now. Confirm that its attitude control system is powered-down." Khambrel talks while exercising the manipulator arm, making certain that all its joints work.

"*Enterprise*, this is Houston. Landsat attitude control system is still active because of an additional problem in the command control receiver on board. But the attitude control system telemetry indicates that the system's gas supply is depleted. There should be no response from its control thrusters when you disturb it with the arm."

"I hope you're right, Houston," Khambrel remarks and turns to Mary. "Secure for possible emergency, Mary. That thing out there may still be active. Dick, are you listening on the net?"

"Commander here. Roger, Khambrel. Do you want me up there?"

"I don't think it's necessary, Dick. But you might stand by in case."

Everyone who flies in space remembers the *Gemini-8* mission where Armstrong and Scott linked up with an *Agena* target satellite, had trouble with their attitude control system, and had to make an emergency landing.

Carefully, Khambrel extends the remote manipulator arm toward the ailing Landsat. "I'm going to slip the claw into the deployment latching mechanism of the Landsat," he explains to Houston and to us. "If I touch it and it starts to tumble because it really is dead, I'll never be able to grab it."

He maneuvers the arm's wrist joint and slips the claw into an attachment on the Landsat's side.

"Got it!"

Suddenly, it's as though he had a wild bronco on the end of a rope. The Landsat begins to spurt attitude control gas from its thrusters.

"It's still hot!"

"Khambrel, get the claw out of there!" Mary snaps.

"Can't! It pushed against the latching mechanism in the claw and gave us a hard latch. I can't turn loose of it!" Khambrel turns to the intercom. "All hands! Red alert! It's still active!"

More quickly than I thought possible, Richard

Edwards sails up through the hatch from the mid-deck. "Khambrel, let it twist! Don't stress the arm joints, and don't try to fight it with our control thrusters. Let it flip us if it wants to!"

"I'm doing that!" Khambrel snaps back coolly.

"I thought Houston said it had depleted its attitude control gas supply," Edwards comments.

"They did," Khambrel reports. "Said it didn't have any left."

The thrashing Landsat on the end of the arm has caused the *Enterprise* to go into a flipping, skewed roll, a complex motion of end-over-end flight with a roll. It isn't a violent maneuver, but it is strange.

"Okay, that means it's just venting residual gas," Edwards ventures. "That's the gas left in the tanks that was too little to measure. It hasn't got much left, so just let it die. But don't let it hit the *Enterprise!*"

"It's almost dead now," Khambrel says. "Just enough poop to give us trouble!"

"Any damage? Any systems inoperable? Report please," Edwards orders.

"The manipulator arm appears to be all right," answers Khambrel. "It's been stressed, but I don't get any indication that anything's wrong."

"All ship systems okay," Mary reports. "I disabled the attitude control system aboard the *Enterprise* so we've lost platform alignment. But we can get it back."

"Everybody just stay put," Edwards advises calmly. "There's no real emergency here. But I don't want anyone aboard moving around in the ship with this tumbling motion in progress. Just hang on and wait. This isn't a bad tumble, and we can get straightened out in short order once we get that wild pig under control out there. You could get badly disoriented if you tried to move around now."

"It's quit thrusting," Khambrel reports. "Do you want me to bring it in first, Dick, or do you want to kill the tumble first?"

"Try bringing it in slowly, Khambrel. Get it in close first. We could twist that arm if we tried to stop the tumble."

"Roger. I'll get it in as close as I can. Let me try to get it into the cargo bay. If I can't, we'll stop the tumble."

Fortunately, there is no sensation of motion other than the twisting, turning surface of the earth visible through the windows. Working carefully, Khambrel inches the Landsat toward the cargo bay.

"Dick, as I bring it in, it's causing our tumble rate to increase," Khambrel says. This is a natural reaction, the same principle that causes figure skaters to spin faster once they pull in their arms. "We've got to kill the tumble gently at the same time I bring it in."

This delicate maneuvering job takes well over two hours. Khambrel nudges the satellite toward the cargo bay, while Edwards and McQuire, who came up to the flight deck to assist, gently fire the *Enterprise* thrusters to reduce the tumble. It requires the utmost in crew coordination to get that Landsat into the cargo bay without breaking the manipulator arm.

Mary managed to get commands to the Landsat to fold its solar panels; if she hadn't, the two pilots were ready to don their space suits, go out through the air lock on the mid-deck, and cut the panels off. But it wasn't necessary.

Finally Khambrel reports, "It's in the cradle and secured in place. But I can't unlatch the manipulator arm. It's jammed in the satellite latch."

"Can you get the cargo bay doors closed?" Edwards asks.

"Affirmative."

"Okay, close the cargo bay doors. Secure for entry. We're about two hours behind schedule, and we've missed our scheduled landing pass, so we'll catch it on the next one."

On our normal, planned flight schedule, the *Enterprise* would have started entry into the atmosphere west of San Diego, streaked across the United States, and made a straight-in landing at Cape Canaveral. But the recalcitrant Landsat has made us two orbits late. We can still land at the Cape, though, because the *Enterprise* is surprisingly maneuverable. It can land at an airport as much as 1,500 miles to the side of its course.

Edwards, McQuire, and Johnson stop the tumble, get the *Enterprise* moving properly in orbit again, and shut down the on-orbit operations. Radio conversation with Houston confirms that all *Enterprise* systems are operable and ready for entry on the next pass over the United States.

Edwards turns to us, watching from the mid-deck hatch. "Would you like to ride on the flight deck during entry and landing?"

"But aren't Mary and Khambrel going to be busy at their stations up there?" I question.

"Not during entry and landing," the commander points out. "Their jobs are essentially completed now. Dave and I work hard during liftoff and landing; Mary and Khambrel work hard in orbit. Technically, they're supposed to ride up here during entry and landing, but they don't have any control functions, and they've seen it many times. You two haven't." He smiles at our hesitation and adds, "Don't worry. I have the right to stretch the rules written by the troops on the ground every now and then."

I can detect in Dick Edwards's attitude the long-established tradition of the sea and the air: the captain is the final authority, makes the final decisions, and runs the show as he sees fit. All the rules and regulations written won't change that time-honored tradition. As copilot Dave McQuire had remarked earlier, "The people who write the rules aren't necessarily the ones who have to strap their pink bodies into these flying machines and make them work. A person who takes command responsibility has to have some leeway to change the planned program within the general mission objectives."

This is a definite opportunity! Riding on the mid-deck, one can't see much of anything, except for the limited view afforded the person sitting next to that little ten-inch porthole.

"I actually prefer to ride the entry and landing

on the mid-deck," Mary admits. "I have to get the hatch open after landing and roll-out, so I like to ride next to it."

So, after another sunset and sunrise in orbit, we strap ourselves into the two seats behind the pilots on the flight deck. Mary makes certain that our Snoopy helmets are hooked up and that our restraint harnesses are secured properly.

In spite of the trouble with the Landsat, everything is ready for entry.

Edwards and McQuire turn the *Enterprise* around, using the attitude control rocket thrusters. The tail of the ship now points in the direction of our motion in orbit; this will allow the thrust from the OMS rocket motors to slow down the *Enterprise* when the motors are fired as retro-rockets. It all seems so simple to do. Edwards uses the control stick in front of him to control the change of attitude just as he would control an airplane.

Our de-orbit burn takes place over Australia. Edwards and McQuire put instructions into the on-board computer, telling it to instruct the guidance system to establish the proper attitude and to fire the OMS rockets for a definite period of time. This changes the velocity of the *Enterprise* by only a few hundred miles per hour, but that is enough to make the ship begin to spiral down toward the earth's atmosphere. Our entry corridor will be on the other side of the world over the eastern Pacific Ocean, Mexico, and the Gulf of Mexico. Edwards will then turn northeastward to land at the Cape.

After a day in weightlessness, the gentle thrust

of the OMS rockets gives us a tiny part of our normal weight back again for a few minutes. It makes us feel heavy, although the acceleration is only a fraction of earth's gravity.

"Houston, this is *Enterprise.* De-orbit burn complete and on schedule. Everything nominal. Coming home to the car barn."

"Roger, *Enterprise.* Our computer here says everything is copacetic. Go to entry attitude at your discretion. Glad the Landsat affair didn't cause undue trouble."

"Thank you, Houston. We'll set up entry attitude now. As for undue trouble with the Landsat, you should have been up here riding this bird. The Landsat contractor should design so that the attitude control system disables itself when the gas supply drops to a minimum value."

"We'll make a note of that, *Enterprise.*"

The two pilots begin to set up their controls and instruments for entry. The control consoles and instrument panels look complicated, but I suppose they don't seem that way to the pilots. Edwards and McQuire appear able to reach out for a switch or a knob without even looking. I suspect that all their training and experience make the intricate display in front of them as familiar to them as the instrument panels of our cars are to us.

Although there are two human pilots aboard, ready to assume manual control of the *Enterprise* if necessary, our entry into the atmosphere and our landing are made under computer-directed automatic control. This is not very much different from modern jet airliners that have been flying and landing under the control of their automatic systems since 1970.

"Well, Dave, let's give our passengers a thrill with some orbital acrobatics," Edwards remarks.

"Wouldn't that be called astrobatics?" McQuire says, smiling.

Edwards must now turn the *Enterprise* around so that it travels nose-first in orbit. Entry into the atmosphere must be made at a precise orientation. Instead of getting into position by slewing the *Enterprise* around horizontally using the yaw thrusters, the pilot moves his control stick. The *Enterprise*'s nose swings up under the nudge from the thrusters. Over it goes onto its back in a half-loop. Turning his control stick, Edwards rolls the big ship over so that its black bottom is earthward again, and we are flying nose-first in orbit.

"During World War I, that was called an Immelmann Turn," Edwards explains. "Our nose is now forward, so I'll change to computer control so that we get the correct nose-up entry angle." He flips a switch. The *Enterprise* slowly brings its nose up until it is pointed almost forty-five degrees to the horizon outside. The nearly flat bottom of the *Enterprise* is now the surface that will strike the atmosphere. It's covered with those reusable black ceramic tiles to withstand the thousands of degrees of temperature that will be caused by the big ship plunging into the atmosphere at about 17,000 miles per hour.

There is no sensation of losing altitude as the

ship proceeds northeastward across the Pacific Ocean. The first indications of our entry into the atmosphere don't show themselves until almost thirty minutes after de-orbit burn.

"Atmospheric interface," McQuire announces. "Altitude four hundred thousand feet. Two minutes to radio blackout."

A gentle force is slowly pushing us downward into our seats as the big black bottom of the *Enterprise* begins to encounter the denser atmosphere closer to the earth. The force grows in strength. We can now see a chimeral, flickering glow dancing over the windows of the flight deck in front. This is the hypersonic shock wave, a pulse of air piling up ahead of the nose of the *Enterprise* and glowing with the heat generated by the impact of the molecules of the atmosphere. Although we cannot see it, the bottom of the *Enterprise* is beginning to glow red-hot. The pilots are monitoring the ship's skin temperatures with remote instruments located at various points.

The *Enterprise* shudders slightly. Its automatic control system, directed by gyros and the computer, is firing the rocket thrusters of the OMS to keep the orbiter in the proper attitude and moving in the right direction. As the ship gets lower into the atmosphere, the aerodynamic controls, such as the elevons on the wings and the rudder on the tail, will be able to direct the craft.

Early space capsules were like ceramic-coated bullets that merely dropped back to the earth through the atmosphere without generating much lifting force. They were designed and developed on the basis of the known technology of warheads for ballistic missiles. But the *Enterprise* is different. It actually flies back through the atmosphere at more than five times the speed of sound. It's a big glider, with its delta-shaped wings generating lift and its control surfaces keeping it level just like an airplane. The *Enterprise* is under control at all times.

The acceleration force caused by the atmosphere slowly decreasing our speed builds up over a period of twenty minutes to a level of 1.5 g's, making us weigh one and a half times our normal earth weight. This is a great deal less entry acceleration than that experienced by early astronauts in their capsules. But, after a day in the weightlessness of orbit, it feels like more than a mere 1.5 g's.

The glow that flickers and wavers over the pilots' windshields finally begins to diminish in intensity, although there are places on the window frames that gleam red-hot. We have passed through the hottest part of the entry now, and the shock wave is starting to subside.

"Houston, *Enterprise* here. Radio check. How do you read me? Over." Edwards is calling on the radio, checking the communications now that the *Enterprise* has passed through the radio blackout phase of entry. The sheath of superheated air of the hypersonic shock wave prevented radio waves from leaving or reaching the radio antennas on the *Enterprise*. Now, however, radio waves should be getting through the supersonic shock wave, which is much cooler.

"*Enterprise*, this is Houston. We read you loud

and clear. Radar contact, and we show you on track over the Gulf of Mexico. Approximately one minute from dogleg turn to the northeast to pick up the Cape homing beacon."

"*Enterprise*, roger. Please give us the terminal weather forecast for the Cape."

"*Enterprise*, Houston. On the hour, Cape weather two-five thousand broken, one-seven thousand scattered, three thousand overcast, visibility five miles in haze and light rain showers. Winds three one zero at five knots. Temperature seven eight. Altimeter two niner niner three. Cumulus buildups reported by commercial airliners in northeast quadrant. Radar shows no thunderstorm cells near your flight path."

"*Enterprise*, roger. We copy Cape weather and standing by to copy approach clearance."

"*Enterprise*, Houston. Miami Air Traffic Control Center clears *Enterprise* to Dogleg Approach Golf. Set radar transponder seven seven five five for radar identification. You are cleared to final approach fix."

The *Enterprise* carries the same radar transponder equipment as a commercial airliner or military aircraft; this equipment receives a radar signal from the ground and sends back a strong reply with an identifying code of pulses mixed with it. On the ground, the computers at the radar site decode this reply and present a bright target on the air traffic controller's radar screen along with the radar identification number from the transponder. Even though the *Enterprise* is a space vehicle, it now functions like an airplane coming in for an unpowered landing through airspace that is also used by airliners, military aircraft, and private planes.

"Roger, Houston. *Enterprise* copies ATC clearance and squawking seven seven five five. Altitude one hundred twenty thousand feet and commencing left turn to Dogleg Approach Golf."

Gliding in the *Enterprise* is somewhat like being aboard a streamlined brick. The descent angle through the atmosphere is almost twenty-five degrees, while most airliners descend at an angle of three to five degrees. This means that the *Enterprise* loses altitude much faster than airliners and heads down toward the earth at a much sharper angle. The stubby delta wings of the *Enterprise* require that the glide be flown in a nose-high attitude for maximum lift and range.

We're below the speed of sound now and slowly dropping to our approach speed of 300 miles per hour. We can see nothing below us through the windows but a white cloud deck. The lack of visibility doesn't really matter, though, for Edwards and McQuire will be making a typical automatic landing, depending upon the readings of their instruments and standing by to take over from the computer if necessary.

The ride is smooth now, but the earth-normal gravity that has returned makes us feel very heavy after a day in orbit.

The cloud deck below appears to be approaching with great rapidity. It's now evident that we are coming down much more quickly than an ordinary airliner.

"Houston, this is *Enterprise*. We have acquired the Cape landing system signal."

"Roger, *Enterprise*. Houston is handing you off to the Cape approach control now."

"*Enterprise*, this is Cape Approach Control. How do you read?" A new voice comes through the earphones.

"Cape Approach, this is *Enterprise*. We read you loud and clear. How me?"

"Loud and clear, *Enterprise*. We show you on glide path turning to approach heading. Four minutes to touchdown."

We appear to drop right through the broken cloud layer at 25,000 feet. The world suddenly becomes white all around us, and I lose my sense of where the horizon is. Then we whisk through the second layer of clouds at 17,000 feet.

"*Enterprise*, this is Cape Approach. Altitude fifteen thousand. Two minutes from touchdown. We indicate your landing system has acquired and locked."

Edwards and McQuire appear to be impassive statues in their seats, but we know they must be scanning the instrument panels in front of them as the automatic landing system brings the *Enterprise* down through the clouds for an absolutely hands-off landing.

Then we are in the final cloud deck, a complete overcast at 5,000 feet. We break out of it with great suddenness at 3,000 feet—and the flatlands of Florida are below us!

Ahead, over the ship's nose, we can see the long

runway at the Cape. The *Enterprise* is precisely aligned with the runway! At a flying speed of 300 miles per hour, it doesn't take long to get there.

"Cape Approach, this is *Enterprise*. Altitude nine hundred feet. Initial flare beginning. . . . On three-degree glide slope now. . . . Stand by for gear-down. . . ."

The approach end of the runway disappears under the nose of the *Enterprise*.

"Two hundred feet! Gear down!" McQuire's left hand moves slightly to push a button on the instrument panel. Although we cannot see it, the tricycle landing gear drops suddenly from beneath the wings and nose of the *Enterprise*.

"Gear down and locked! We have three green lights! Final landing flare! Over the fence at two hundred!"

Up and up comes the *Enterprise*'s nose until it is pointed toward the overcast sky above. The falling feeling we have had now disappears as the *Enterprise* begins to level off for the final landing maneuver.

In a moment, we feel the main landing gear wheels touch the runway.

"Pretty good, boss," McQuire comments to the pilot. "Three feet per second at touchdown."

"With a little help from some electronics," Edwards replies laconically.

The nose begins to drop slowly, and it drops . . . and drops . . . and drops until we can see the horizon and the rest of the runway ahead. Finally, the nose landing gear touches. McQuire's left hand

The artist's cutaway drawing shows Mary Howard, Khambrel Johnson, and the two of us at stations on the flight deck while we are running our experiment in space.

Manipulator arm extended, the *Enterprise* proceeds to retrieve a Landsat from orbit because it has stopped working. The inoperative satellite will be taken back to earth for repairs in the *Enterprise*'s cargo bay. It will be repositioned in orbit on another Shuttle flight.

Johnson, Edwards, and McQuire work together to maneuver the faulty Landsat into the *Enterprise*'s cargo bay while controlling the orbiter's tumble rate. Through the aft windows on the flight deck, the Landsat can be seen attached to the ship's manipulator arm.

Its belly glowing red-hot from friction with earth's upper atmosphere, the *Enterprise* begins its entry over the Pacific Ocean. The special ceramic tiles that cover the orbiter's critical surfaces absorb the heat and keep its aluminum and titanium structure from melting during entry.

After the blazing entry into earth's atmosphere, the *Enterprise* flies like a glider, maneuvering under the control of its aerodynamic surfaces, such as the elevons on the delta wings and the rudder on the tail. The *Enterprise* can glide as far as 1,500 miles to either side of its entry course.

The *Enterprise* descends steeply at 300 miles per hour on its final landing approach to the 15,000-foot runway at Cape Canaveral. The Vertical Assembly Building is to the right. The landing is made under completely automatic control, but the pilots can take over with manual controls if necessary.

Landing gear down, the *Enterprise* makes its final landing flare before touchdown on the runway at Cape Canaveral, its speed now 200 miles per hour. It lands nose-high like a *Concorde* SST.

After landing, the *Enterprise* will be towed off the runway to the Vertical Assembly Building, where its cargo will be removed from the cargo bay. In two weeks, after minor repairs and maintenance, it will fly into orbit again.

moves quickly to a button on the panel over the windshield, the button that will open the rudder surface as an air brake. Edwards is operating the wheel brakes.

The *Enterprise* rolls and rolls and finally slows to a brake-chattering stop with thousands of feet of runway still in front. We are almost immediately surrounded by vehicles on the runway outside.

"Cape Approach, this is *Enterprise*. On the deck and roll-out complete. Initiating systems shutdown and crew egress." Both pilots are now busy with the post-landing checklists, shutting down various *Enterprise* systems one by one and making the ship safe for the ground crews to approach.

McQuire turns to me, "Okay, you can go downstairs and egress as soon as Mary gets the hatch open."

"Nice landing. In fact, nice flight," I tell him as we unstrap.

"Piece of cake," is the copilot's only comment as he returns to the final checklists.

Mary has the access ladder between the flight deck and the mid-deck in position, and we crawl down. Khambrel is stowing the passenger seats.

Mary opens the hatch in the side of the cabin with one quick turn of the lever in the middle of the door. Quick-opening hatches have been standard in space flight ever since the tragic *Apollo-1* fire that killed three astronauts on the launch pad in 1967. The hatch slowly lowers to a horizontal position, and we smell the warm, damp air of the Cape. Mary drops a thermal protective mat over the edges of the hatch because the exterior of the *Enterprise* is still hot from its passage through the upper atmosphere; it could burn us if we touched it as we get out. We can smell the heat, too—a scorched hot-brick smell from the thermal protective tiles of the *Enterprise*.

"Howdy, astronauts," a new voice calls as the platform of a vertical-lift-bed truck is elevated to the level of the hatch. It's a ground crewman, ready to help us debark from the ship. In NASA terminology, he's the "crew egress technician."

One at a time, we duck out of the cabin opening and step off the hatch onto the truck. There's a ladder on the side of the truck so we can clamber down to Mother Earth again.

Which we do. Solid ground with 1-g acceleration seems strange after a day in orbit, but we'll get used to it again within minutes.

Above us, the *Enterprise* bulks stationary on the runway. It's hard to believe that less than thirty minutes before, it was hurtling through space at 17,000 miles per hour. The ship is far from silent, however, because it's still cooling off. It creaks and groans as if it were complaining about being on earth and asking to be sent back into its natural element, space.

It is just sundown at the Cape, but we can't see the sunset through the overcast. But what matter? We could have seen eighteen sunsets in the last twenty-four hours! And what is a sunset on earth compared to the brilliant spectacle of one in space?

The *Enterprise*, of course, will see many more

sunsets in space, for its career is just beginning. It has been designed to make at least a hundred flights into space. And, as space technology advances even more rapidly because of what the *Enterprise* and its sister ships will permit people to learn, it will probably be making trips into orbit until the year 2000. By that time, there will be newer and more advanced Space Shuttles, of course. But the *Enterprise* is a space truck, and it will keep on trucking into space for years to come.

Our flight is over, but our work has just begun. It will be several hours before the *Enterprise* is towed back to the VAB, where its cargo door can be opened with the special rigging and the payloads can be removed. We'll then get the chance to examine our space wood samples. And the faulty Landsat will be removed and transported to the factory for repair and refurbishment.

Our flight aboard the *Enterprise* has been a great experience. Ten years ago, it would have been an impossible experience for ordinary people like ourselves. Today we can do it, and we are unique. But it won't stay that way. Out there in space, the future is waiting for the human race. Ten years from now, thousands of people will be living in space. Eventually, there will be space settlements and lunar outposts. After all, a whole solar system exists out there, waiting for us to explore it and use it and live in it.

And beyond that, someday, are the stars.

APPENDIX 1

Space Shuttle Characteristics

(values are approximate)

LENGTH

System: 56.14 meters (184.2 feet)
Orbiter: 37.24 meters (122.2 feet)

HEIGHT

System: 23.34 meters (76.6 feet)
Orbiter: 17.27 meters (56.67 feet)

ORBITER WINGSPAN

23.79 meters (78.06 feet)

WEIGHT

Liftoff: 1,995,840 kilograms (4.4 million pounds)
Orbiter landing: 84,778 kilograms (187,000 pounds)

THRUST

Solid Rocket Boosters (total of two): 12,899,200 newtons (2.9 million pounds) each
Space Shuttle Main Engines (total of three): 1,668,000 newtons (375,000 pounds) each
Orbital Maneuvering Engines (total of two): 26,688 newtons (6,000 pounds) each

EXTERNAL TANK

Length: 47 meters (154.2 feet)
Diameter: 8.38 meters (27.5 feet)

SOLID ROCKET BOOSTERS

Length: 45.46 meters (149.16 feet)

Diameter: 3.70 meters (12.16 feet)

CARGO BAY

Length: 18.28 meters (60 feet)

Diameter: 4.57 meters (15 feet)

PAYLOAD CAPABILITIES

Carry into orbit: 29,484 kilograms (65,000 pounds) of cargo plus seven people

Return from orbit: 14,515 kilograms (32,000 pounds) of cargo plus seven people

REUSABILITY

Designed for a minimum of 100 missions into orbit

APPENDIX 2

Space Shuttle Jobs

The Space Shuttle provides a degree of flexibility never before attained in space operations. It permits us to treat outer space as a resource to be utilized, not just a mystery to be tested and explored. Because the Space Shuttle provides routine access to earth orbit quickly, on short notice, and at greatly reduced costs, it can help us perform many beneficial jobs in space—carrying earth resources satellites, communications satellites, and weather satellites to and from orbit as well as allowing the manufacture of many useful new products and space structures. With the Space Shuttle, the repair and replacement of unmanned satellites in orbit and the continuous manned operation of space stations are everyday practices, resulting in benefits in such areas as personal communications, pollution monitoring, resources management, energy production, and national security as well as continued scientific and space exploration.

During the 1980 decade, the Space Shuttle can carry into orbit both manned and automatic payloads and satellites for the following earth-oriented activities.

Production and Management of Food and Forestry Resources

Global crop production forecasting
Water availability forecasting

Land use and environmental assessment
Living marine resource assessment
Timber inventory
Rangeland assessment

Prediction and Protection of the Environment

Large-scale weather forecasting
Weather modification experiments
Climate prediction
Stratospheric changes and effects
Water quality monitoring
Global marine weather forecasting

Protection of Life and Property

Local weather and severe storm warnings
Tropospheric pollutants monitoring
Hazard forecasting from on-site measurements
Communications and navigation
Earthquake prediction
Control of harmful insects

Energy and Mineral Exploration

Solar power stations in space
Power relay via satellites
Hazardous waste disposal in space
World geologic atlas

Transfer of Information

Domestic communications
Intercontinental communications
Personal communications worldwide

Use of Space Environment for Scientific and Commercial Purposes

- Basic studies of physics and chemistry
- Materials science and technology
- Commercial inorganic materials processing
- Biological materials research, application, and processing
- Effects of gravity upon terrestrial life forms
- Living and working in space
- Physiology and human disease processes research

Earth Science

- Studies of the earth's magnetic field
- Earth crustal dynamics studies
- Ocean interior and dynamics studies
- Dynamics and energy balances in the lower atmosphere
- Structure, chemistry, and dynamics of the upper atmosphere
- Interactions between the lower and upper atmosphere

APPENDIX 3
Space Shuttle Fares

How much does it cost to fly on the Space Shuttle? The answer is anywhere from about $3,000 to more than $21 million, depending on how much the cargo weighs, how much volume is involved, whether it's on a reservation or stand-by basis, and what optional services are desired.

At the low end of the fare structure is the Quick Getaway Special, which permits an individual or an organization such as a school to fly a payload in the Shuttle for $10,000 or less. These payloads are flown on a space-available basis, tucked into otherwise empty corners of the cargo bay. To qualify for the low fare, the proposed payload must involve scientific research to be carried out in orbit, must weigh less than 200 pounds, must have a volume of less than five cubic feet, and must be completely self-contained. Any additional services such as electrical power from the Shuttle cost extra.

It costs only $500 to reserve space for a Quick Getaway Special, most of which is refundable if you later decide not to fly your payload. Many schools, colleges, universities, research institutes, and small businesses have reserved almost a hundred Quick Getaway Specials.

The full $21-million fare is for nongovernment customers using the full thirty-two-ton payload capacity of the Space Shuttle on a firm reservation basis.

In between the Quick Getaway Special and the full $21-million flight are a range of fares for customers sharing a Shuttle flight with other customers, customers willing to fly on a stand-by basis, and customers who have been involved in the development of the Spacelab, which is carried in the cargo bay of some Shuttle flights. There are reduced prices for payloads having exceptional merit scientifically or technically. Lower rates are also offered for stand-by payloads, for optional launch dates rather than fixed dates, and for reserved space on future missions. It is NASA's policy to allow several users to share a Shuttle flight, with the price of each user's share being a fraction of the cost of the total mission. Charges for shared flights are based on the size of the payload and its destination in space.

A Space Shuttle customer may also sell part of his payload reservation space to others.

The Space Shuttle orbiter can carry as many as seven people per flight. Three of these are crew members supplied by NASA—the pilot, the copilot, and the mission specialist. The payload specialist may also be a NASA person.

Besides supplying the Shuttle crew, NASA provides such standard services as experiment design reviews, safety reviews, flight planning, and transmission of payload data back to earth by radio. Optional services available at additional cost include special crew training, upper stages, special mission kits, revisit and retrieval of satellites, and special equipment. Additional charges are also made for short-term reservations, postponements, and cancellations.

NASA reviews its Shuttle fare structures regularly to insure that decreased costs due to experience and increased costs due to inflation are taken into account.

Anybody can fly on the Space Shuttle. For more information on how to arrange it, write to the Operations Director, Space Transportation Systems, NASA Headquarters, 400 Maryland Avenue S. W., Washington, D. C. 20560.

GLOSSARY

Acceleration A change in velocity that produces a gravitylike force.

Acquired To receive a radio signal or radar reply from an object.

Air lock A passage between a pressurized and an unpressurized volume of a spacecraft, with a door, or hatch, opening to each volume. To go from pressurized volume to unpressurized volume, one enters the air lock, closes the pressure-side door, activates controls that lower the air-lock pressure to zero, then opens the door to the unpressurized volume. This prevents loss of valuable atmospheric pressure. To reenter pressurized volume, the procedure is reversed.

Apogee That point in the orbit of an object around the earth when it is at its farthest orbital distance from the earth.

ATC clearance Air Traffic Control permission for an airborne vehicle to enter a volume of airspace or to fly on a given course through a volume of airspace.

Atmospheric interface That point in the entry of a spacecraft into the earth's atmosphere when the first effects of the atmosphere are detectable.

Attitude The orientation of an aircraft or spacecraft with respect to a given frame of reference such as the local vertical direction.

Automatic landing mode The operation of a Space Shuttle's automatic, computer-controlled guidance and control system to bring about a fully automatic landing of the orbiter.

Automatic pilot (autopilot) A computer-controlled electromechanical device that keeps a ship, airplane, or space vehicle pointed in a predetermined direction and moving along a predetermined course.

Automatic sequencer A device that performs a series of operations in a predetermined sequence without human control once the device has been started.

Beta cloth A fireproof cloth made of glass fibers and other nonflammable substances and used to make space-going garments.

Bulkhead Nautical and astronautical term for a wall.

Burn A thrust period of a rocket motor.

Burnout The cessation of rocket thrust, caused either by deliberately shutting off the flow of rocket propellants or by letting a rocket motor operate until its propellant supplies are exhausted.

Bus voltages The electrical voltage on the main electrical distribution wiring of a spacecraft.

Canopus A very bright star in the southern sky used to orient spacecraft because its brilliance makes it easy to locate.

Console An instrument panel, usually a large one, in which all control and measurement functions are grouped together.

Copacetic A slang word, origin unknown, meaning excellent.

Crew egress NASA terminology for the debarking of the flight crew from an orbiter after landing.

Crew ingress NASA terminology for the boarding of the flight crew in an orbiter.

Deck The floor of a ship or space vehicle. Although there is no "up" or "down" in weightlessness, the deck is still used as the "down" reference. The term also refers to floor levels such as flight deck, mid-deck, and lower deck.

Delta wing An aircraft wing with a plane shape like a triangle or the Greek letter "delta."

De-orbit burn The rocket thrust period that reduces a spacecraft's orbital speed and permits it to change orbits.

DISKY Astronaut shorthand terminology for the computer keyboard used to give instructions to an on-board computer in a spacecraft.

Disorient To become confused about one's position or attitude in space.

Dogleg turn A turn to a new direction during the landing of an aircraft or spacecraft.

Entry The encounter of a space vehicle with the upper atmosphere of the earth; was called reentry in the early days of space flight.

External Tank (ET) The aluminum tank 154 feet long and 28 feet in diameter in which are stored the propellants—liquid hydrogen and liquid oxygen—used by an orbiter's rocket engines. The External Tank is dropped from the orbiter before orbit is attained, and is allowed to impact in the Indian Ocean; it is the only part of the Space Shuttle system that is expended on each flight.

Final pad The final computer program and instructions computed by Launch Control and radioed to a spacecraft's flight crew in the form of a long list of numbers that is written on a pad of paper and finally entered into the on-board computer by a flight crew member using a DISKY.

First-day covers Envelopes bearing postage stamps issued on the first day the stamps become available, or stamped envelopes taken on a unique and specific voyage.

Flame trench A large concrete pit or ditch under the launch pad of a spacecraft, serving to direct the rocket exhaust away from the craft and the launch equipment during liftoff.

Flare The action of an orbiter leveling out to make the final descent before landing.

G A unit of force exerted on a body by gravity; 1 g is an acceleration force equal to the force of earth's gravity; 2 g's is an acceleration force twice that of earth's gravity, making a body weigh twice as much as normal.

Glide slope The final landing approach path of a Space Shuttle orbiter during its landing on a runway; it is usually defined by the direction of the landing system's radio beam.

Hypergolic propellants Liquid rocket propellants—a fuel and an oxidizer—that ignite spontaneously upon contact with each other and do not require any type of igniter; nitrogen tetroxide and monomethylhydrazine are used in the Space Shuttles.

Hypersonic shock waves A sharp buildup of air molecules in front of an aircraft or spacecraft moving through the earth's atmosphere at a speed exceeding five times the speed of sound (Mach-5), producing a bow wave similar to that of a motorboat running at high speed on water.

Landsat Shorthand term applied to Earth Resources Surveying Satellites.

Laser A device capable of producing a highly intense beam of light energy. Basically, chemical, electrical, or mechanical energy is forced or "pumped" into a laser, which then converts this energy into a very concentrated beam of infrared or visible light of a single frequency where all light vibrations are coherent, or in step with each other.

LOX Shorthand term for *l*iquid *ox*ygen.

Manifest A detailed, item-by-item, computer-generated list of all things that are aboard an orbiter.

Mass A basic property of matter that produces the sensation of weight when accelerated by gravity or by a rocket motor.

Mission The flight of a space vehicle including all the operations that must be performed during the flight.

Mobile launcher A movable launch pad; each Space Shuttle is assembled in the Vertical Assembly Building on a mobile pad that is moved to the launch area for liftoff.

Module A component of a system, large or small, that can be removed from the system as a single unit and replaced as a single unit.

Momentum A basic property of matter that makes an object difficult to put into motion and difficult to stop once it is moving.

Nitrogen purge The flooding of a system or spacecraft volume with nitrogen, a dry gas that does not react with most materials used in rocket and space vehicles, in order to purge, or remove, all air or other chemicals from the system.

Orbital Maneuvering System (OMS) A system of several small liquid propellant rocket engines fueled by hypergolic propellants and used to maneuver an orbiter in space.

Orbiter The winged portion of the NASA Space Shuttle system that carries crew and cargo and returns to earth to land like an airplane on a runway.

Orient To give direction to one's self or to an object in space.

Pitch The action of a ship, airplane, or space vehicle when it rotates in a nose-up or nose-down direction.

Platform alignment The basic frame of reference of a gyro-stabilized guidance system is the "platform" on which it is mounted; the platform in turn must be aligned with some external frame of reference such as a very bright star whose position in space is known.

Powered-down Astronautical term for shutting off or turning off a device or system.

Power-up The act of putting electrical, hydraulic, or gas pressure to a device or system; literally, turning on the power.

Radar transponder A radio receiver-transmitter in an airplane or spacecraft that detects a signal from a ground-based radar set and sends back a strong signal, enabling the craft to appear as a bright target on a radar screen.

Radio blackout The cessation of radio communications that occurs when an orbiting object enters the earth's atmosphere at high speeds, creating an incandescent shock wave ahead of it that radio waves cannot penetrate.

Readouts The digital or dial readings of instruments. In an automobile, the speedometer dial provides readouts of speed, and the digital odometer provides readouts of distance traveled.

Reduced data Usable information drawn from raw data, the data that originally comes from an experiment or test; the act of converting raw data into information one can use is called reducing the data.

Refurbishment The repair, maintenance, replacement of parts, and general rebuilding of a device.

Rendezvous A planned meeting of two space vehicles in orbit.

Retro-rockets Rocket motors that produce thrust opposite to the direction of a space vehicle and that therefore slow down the vehicle.

Roll The action of a ship, airplane, or space vehicle when it rotates around its fore and aft axis.

Roll-out After an airplane or a Space Shuttle orbiter touches down on a landing runway, it "rolls out" along the runway, losing speed through braking action applied by the pilots in the cockpit.

Runway threshold The end of the landing runway over which an airplane or orbiter must pass on its final landing maneuver.

Service towers Steel frameworks that are wheeled up to a spacecraft to permit people to work on the craft while it is in the prelaunch vertical position. The towers are usually wheeled away from the spacecraft shortly before launch time.

Simulator A ground-based replica of a Space Shuttle flight deck and mid-deck that exactly duplicates the parts of a real orbiter; some simulators are combined with computers so that it is possible to "fly" or otherwise operate them in an exact training version of a particular mission.

Snoopy helmets Black-and-white cloth helmets holding earphones and microphones and worn by space crew members; so-called because they resemble the appearance of a well-known comic strip character who was adopted by astronauts before the first manned landing on the moon.

Solid Rocket Boosters (SRBs) Two recoverable solid propellant booster rockets, each 149 feet long and 12 feet in diameter, used to boost a Space Shuttle system to an altitude of about twenty-five miles, where they are detached, parachuted into the Atlantic Ocean, and recovered for reuse.

Space Shuttle Main Engines (SSMEs) Three liquid propellant rocket engine thrust chambers, each producing 375,000 pounds of thrust, that are mounted on the rear of an orbiter, drawing their propellants (liquid hydrogen and liquid oxygen) from the External Tank.

Squawking Aerospace term for setting the proper identification numbers into a radar transponder.

Stack Astronautical term referring to the completely assembled launch vehicle; originated in the days when space vehicles were multi-staged rockets in which various stages were placed, or stacked, atop one another.

Star tracker An automatic device that seeks out a predetermined star in the sky and stays pointed at that star until instructed to do otherwise. A star tracker therefore can give a guidance system or gyro-stabilized platform its necessary frame of reference.

Supersonic shock wave A buildup of air molecules ahead of an aircraft or spacecraft similar to a hypersonic shock wave, but occurring at speeds between Mach-1 and Mach-5. Because of the lower speed, the physical characteristics of a supersonic shock wave are different from those of a hypersonic shock wave.

Surface tension The physical characteristic of a liquid at its interface with another liquid or with a gas such as air; the surface tension of water, for example, causes it to "wet," or surround, anything with which it comes into contact.

Telemetry Literally, measurement at a distance; the use of radio communications to send information concerning the operation of a device and the extent of the operation.

Thrust The force produced by the operation of a rocket engine.

Topping The final filling of a propellant tank to its maximum capacity just before liftoff.

Torque The twisting force exerted by a turning shaft.

Transient crew quarters A place to eat, sleep, and

live for Space Shuttle crew members who come to Cape Canaveral to fly a mission and do not live at the Cape.

Translate To move sideways, up, or down in space without changing the attitude of the spacecraft.

Vernier engine A small rocket engine in an orbiter's OMS that produces very little thrust and is used to perform very delicate, small maneuvers in space.

Vertical-lift bed A type of truck having a horizontal platform that can be raised vertically with hydraulic pistons to various levels.

Whiskers Long (one-inch), very fine filaments of metallic crystals grown on earth or in space; whiskers are the strongest materials known, and those grown in space are longer and stronger than those grown on earth.

INDEX

Numbers in italics refer to diagrams on pages 8–11

About the Author

G. HARRY STINE has been writing about space and working with rockets and space vehicles since 1951. *Shuttle into Space* is his sixteenth book on rocketry and astronautics.

Many people know Mr. Stine as the founder of the space-age hobby of model rocketry and as the author of *Handbook of Model Rocketry,* the primary reference book for the sport. However, he is also a futurist and one of the pioneers in space industrialization. He is a Fellow of the Explorers Club in New York and the British Interplanetary Society in London. (He was nominated for fellowship in 1952 by the famed science-fiction writer Arthur C. Clarke.) As an Associate Fellow of the American Institute of Aeronautics and Astronautics, he is a member of the technical committees on space processing and aerospace history for this organization of professional engineers.

Although Mr. Stine will probably never be able to pilot a Space Shuttle, he comes as close to it as he can by flying his own private airplane whenever possible. And he still builds and flies model rockets. He lives in Phoenix, Arizona, with his wife, a son, two grown daughters, two golden retrievers, and a cat with twenty-four claws.